1 $ 1 4 9 5 L 2

THE CHEMISTRY OF
BUILDING MATERIALS

THE CHEMISTRY OF BUILDING MATERIALS

R. M. E. DIAMANT, MSc, DipChemE, AMInstF
Lecturer in Chemistry and Applied Chemistry
University of Salford

Applied Chemistry Series
Number 1
General Editor
Professor G. R. Ramage, University of Salford

BUSINESS BOOKS LIMITED
London

First published 1970

© RUDOLPH MAXIMILIAN EUGEN DIAMANT 1970

SBN 220 79933 4

This book has been set 10 on 12 pt Times Roman and printed in England by C. Tinling & Co. Ltd., for the publishers, Business Books Limited (Registered office: 180 Fleet Street, London EC4) Publishing offices: Mercury House, Waterloo Road, London SE1

MADE AND PRINTED IN GREAT BRITAIN

Contents

List of Figures

Preface

With the increasing sophistication of building methods it is becoming more important for all concerned with the building industry, the structural engineering industry and the civil engineering field to know more about the basic chemical and physical properties of the many materials used today.

This book is intended to be both a student's textbook for Universities and Colleges of Technology, and a handbook for the various professions concerned with building.

It helps if readers have a basic training in chemistry of about Ordinary Level GCE standard, but this is not absolutely essential provided an intelligent approach is made, and there is some willingness to read round topics which require a certain amount of basic knowledge.

The units used throughout this book are standard SI (Systeme International d'Unites) as recommended by the ISO in its meeting at Elsinore in June 1966, and in document 68/25097 of the British Standards Institution: 'The Use of the Metric System in the Construction Industry'. Many practising engineers may find these somewhat unfamiliar, although in fact after very short practice they are indeed much easier to work in than the more traditional units. For example, the traditional form of expressing a thermal conductivity figure is Btu.in/hr.ft².degF which is far clumsier than the SI unit of W/m.degC (W/m.degK), and also much more difficult to work in. SI units are being adopted throughout the world with the present exception of the United States and Canada. Readers who are not familiar with these units should consult: *Understanding SI Metrication*, Angus and Robertson, London (1969).

However, a complete set of conversion factors is included to enable the text to be correlated with work published elsewhere in the traditional units.

The author would like to record his appreciation to the editor of this series of books, Professor G. R. Ramage, Chairman of the Department of Chemistry and Applied Chemistry, the University of Salford, as well as to the many firms and trade associations who have supplied him with technical information and diagrams.

R. M. E. DIAMANT
University of Salford

SI Conversion Factors

Length
1 yard=0·914 4 metre
1 foot=0·304 8 metre
1 inch=25·4 millimetre

Area
1 yard^2=0·836 127 metre^2
1 foot^2=0·092 903 metre^2
1 inch^2=645·16 millimetre^2

Volume
1 yard^3=0·764 555 metre^3
1 foot^3=28·316 8 decimetre^3 (litre)
1 inch^3=16·387 1 centimetre^3

Capacity
1 UK gallon=4·546 09 decimetre^3 (litre)
1 US gallon=3·785 41 decimetre^3 (litre)

Mass
1 UK ton=1,016·05 kilogramme
1 US ton=907·184 4 kilogramme
1 hundredweight (UK)=50·802 3 kilogramme
1 pound=0·453 592 37 kilogramme
1 ounce=28·349 5 gramme
1 grain=0·064 798 9 gramme

Temperature
°F=9/5°C+32
K=°C+273·15
°R=1·8K

Heat
1 British thermal unit=1·055 06 kilojoule
1 centigrade heat unit=1·899 108 kilojoule
1 kilocalorie=4·186 8 kilojoule

Power
1 horsepower=0·745 7 kilowatt
1 horsepower (metric)=0·735 499 kilowatt

Pressure
1 atmosphere=101·325 kilonewton/metre^2
1 inch of mercury=3·386 389 kilonewton/metre^2
1 centimetre of mercury=1·333 22 kilonewton/metre^2
1 bar=10^5 newton/metre^2
1 inch water gauge=249·082 newton/metre^2
1 centimetre water gauge=98·063 8 newton/metre^2
1 pound (f)/foot^2=1·488 164 newton/metre^2
1 pound (f)/inch^2=6·894 76 kilonewton/metre^2
1 ton (f) (UK)/foot^2=107·252 kilonewton/metre^2
1 ton (f) (UK)/inch^2=15·444 3 meganewton/metre^2=15·444 3 newton/millimetre^2

Density   1 ton (UK)/yard³ = 1·328 94 tonne/metre³
1 pound/foot³ = 16·018 5 kilogramme/metre³
1 pound/inch³ = 27·679 9 kilogramme/decimetre³

Concentration 1 grain/100 foot³ = 0·022 883 5 gramme/metre³
1 ounce/gallon (UK) = 6·236 gramme/decimetre³
1 grain/gallon (UK) = 14·254 gramme/metre³

Thermal conductivity

1 Btu/foot hour degF = 1·730 73 watt/metre degC
1 Btu inch/foot² hour degF = 0·144 228 watt/metre degC
1 kcalorie/metre hour degC = 1·163 watt/metre degC

Thermal conductance

1 Btu/foot² hour degF = 5·678 26 watt/metre² degC
1 kilocalorie/metre² hour degC = 1·163 watt/metre² degC

Moisture and air diffusivity

$$1\ \text{grain inch/foot}^2\ \text{inch mercury hour} = 1\cdot453\,\frac{\text{milligramme metre}}{\text{meganewton second}}$$

$$= 5\cdot231\,6\,\frac{\text{gramme metre}}{\text{meganewton hour}}$$

Note: Thermal conductivities can be expressed as either watt/metre degC or watt/metre degK.

Basic SI Units

Physical quantity	Name of unit	Symbol for unit
Length	metre	m
Mass	kilogramme	kg
Time	second	s
Electric current	ampere	A
Thermodynamic temperature	kelvin	K
Luminous intensity	candela	cd

Symbols for units do not take a plural form

Supplementary Units

Physical quantity	Name of unit	Symbol for unit	Definition of unit
Plane angle	radian	rad	} Dimensionless
Solid angle	steradian	sr	
Energy	joule	J	$kg\ m^2\ s^{-2}$
Force	newton	N	$kg\ m\ s^{-2} = J\ m^{-1}$
Power	watt	W	$kg\ m^2\ s^{-3} = J\ s^{-1}$
Electric charge	coulomb	C	$A\ s$
Electric potential difference	volt	V	$kg\ m^2\ s^{-3}\ A^{-1} = J\ A^{-1}\ s^{-1}$
Electric resistance	ohm	Ω	$kg\ m^2\ s^{-3}\ A^{-2} = V\ A^{-1}$
Electric capacitance	farad	F	$A^2\ s^4\ kg^{-1}\ m^{-2} = A\ s\ V^{-1}$
Magnetic flux	weber	Wb	$kg\ m^2\ s^{-2}\ A^{-1} = V\ s$
Inductance	henry	H	$kg\ m^2\ s^{-2}\ A^{-2} = V\ s\ A^{-1}$
Magnetic flux density	tesla	T	$kg\ s^{-2}\ A^{-1} = V\ s\ m^{-2}$
Luminous flux	lumen	lm	$cd\ sr$
Illumination	lux	lx	$cd\ sr\ m^{-2}$
Frequency	hertz	Hz	cycle per second
Customary temperature, t	degree Celsius	°C	$t/°C = T/K - 273 \cdot 15$

Fractions and Multiples

Fraction	Prefix	Symbol	Multiple	Prefix	Symbol
10^{-1}	deci	d	10	deka	da
10^{-2}	centi	c	10^2	hecto	h
10^{-3}	milli	m	10^3	kilo	k
10^{-6}	micro	μ	10^6	mega	M
10^{-9}	nano	n	10^9	giga	G
10^{-12}	pico	p	10^{12}	tera	T
10^{-15}	femto	f			
10^{-18}	atto	a			

Chapter One **Cements and Concrete**

Although the name 'cement' is given to any substance capable of uniting other materials to form a solid whole, this chapter is restricting the term to substances which are technically known as 'calcareous cements', i.e. cements containing calcium.

By far the biggest output of cement today is in the form of *Portland cement*, but there is also quite a number of other types in use, which have more specialized outlets and are not marketed in such large quantities.

Concrete is the term that is used to describe an aqueous mix of cement with sand and crushed mineral matter. The sand is an essential ingredient as it is required to react with calcium hydroxide liberated during the setting of the concrete. In general, siliceous material that has a diameter of less than 5 mm is called *sand*, while particles with a larger size are called *gravel*. Crushed rocks are also usually added to concrete mixes, the main types being granite, basalt, sandstone, quartzite and limestone.

Blast furnace slag, which is a mixture of lime, alumina and silica with small quantities of other materials such as oxides of magnesium, iron, manganese and even titanium, is often added to concrete mixes. Flyash, which is obtained in enormous quantities from coal-fired power stations and other furnace installations is also often added to cement in order to cheapen the product and also to modify its properties.

Cements hold the aggregates together by virtue of the fact that a surface chemical reaction takes place between the free calcium oxide of the cement and the acidic SiO_2 contained in the aggregate. A number of light-weight aggregates is also used in concrete mixes but these will be dealt with in Chapter 2.

Concrete is good in compression but very poor in tension. Because of this, it is usually reinforced with steel bars or mesh. A surface chemical reaction takes place between the concrete constituents and iron oxide films on the surface of the reinforcement irons. For this reason reinforcement irons should not be brightly polished when they are used, but rather somewhat oxidized.

B

Additional methods of reinforcing concrete to counteract its poor tensile properties are the techniques of post-tensioning and prestressing, but these are important fields of study in their own right and will therefore not be dealt with further in this book which is concerned with the chemistry of the material only. Even bamboo is used to reinforce concrete, a practice which is quite common in the Far East.

When cement and its aggregates are mixed with water, a material of a plastic constitution is produced. This cement mix is then poured into moulds and begins to *set* immediately. During the setting process concrete is usually vibrated to eliminate any air pockets and to obtain consolidation of the cement mix and aggregate. After the setting process has proceeded for a certain time, *hardening* begins. With most concretes there is a very considerable passage of time between the stage when the concrete has set and its achievement of final hardness. But there are considerable differences in both setting times and hardening times with different types of cements. Concrete can be hardened a good deal quicker by heat-curing. This is usually carried out by exposing the set concrete to steam. Superheated steam in autoclaves is often used for the manufacture of concrete products.

1.1 Portland cements

Portland cement is produced by heating lime-bearing material with material containing silica, alumina and some iron oxide. The first stage is the careful compounding of the ingredients, followed by heating almost to melting-point in various kinds of kilns under oxidizing conditions. The product, which is in the form of vitrified pellets, is called *clinker*. It is ground together with gypsum ($CaSO_4 . 2H_2O$) and other materials, which act as retarders, to produce cement powder. The properties of the cement are very closely governed by the nature of the constituents of the clinker and above all, by the balance of these constituents with each other. There is quite a number of different Portland cements on the market, each with their own specific properties.

The main types are the following:

1 *Normal or standard Portland cement*, which has average values for initial and final hardening. It is the main type made.

2 *Rapid-hardening Portland cement* is ground rather finer than the normal variety, although its composition only differs very slightly from it. It sets in about the same time, but reaches its final strength more rapidly.

3 *Quick-setting Portland cement* sets in a short time, but its hardening time is similar to that of normal Portland cement.

4 *White Portland cement* is used for facade purposes. Its pale colour is due to the fact that the percentage of iron oxide in its composition is drastically restricted.

5 *Waterproofed Portland cement* is normal Portland cement mixed with small quantities of calcium stearate $[Ca(C_{17}H_{35}COO)_2]$ or small quantities of mineral oil.

6 *Low-heat Portland cement* has a restricted percentage of aluminates, so that the heat given out during setting is much less than with other grades.

7 *Sulphate-resistant Portland cement* resists attack by sulphates in water and is used for casting sections liable to be in permanent contact with ground waters.

8 *Ferrari cement* contains a high percentage of iron oxide, has good chemical resistance and is fairly cheap.

9 *Air-entrained Portland cements* usually employ normal cement, or cement with a high proportion of blast furnace slags to which an air-entraining agent has been added. The most common air entraining agents are alkali salts of wood resins, calcium lignosulphate, sodium salts of cycloparaffin carboxylic acid such as sodium cyclohexanoic acid, calcium salts of proteins obtained during the treatment of animal hides, sodium lauryl sulphates and many others. The quantity of such agents added varies with its nature, but is usually of the order of 0·1—0·5%.

THE CEMENT NOTATION

The chemical formulae used in cement chemistry are usually written down as oxides, because this is much simpler than to attempt to write down the macro-molecular formula of the compound actually formed. For example, if a cement contains $Ca_{3n}Si_nO_{5n}$ molecules where n is a very large number, it is probably much more useful to write down its formula simply as $3CaO \cdot SiO_2$. However, because of the complexities of many cement formulations, a further simplification is employed which is called the *cement notation*. The abbreviations commonly used are

$C = CaO$	$F = Fe_2O_3$	$A = Al_2O_3$
$S = SiO_2$	$M = MgO$	$H = H_2O$
$N = Na_2O$	$K = K_2O$	$L = Li_2O$
$P = P_2O_5$	$f = FeO$	$T = TiO_2$
$\bar{S} = SO_3$	$\bar{C} = CO_2$	

For example, the compound $6CaO \cdot Al_2O_3 \cdot 3SO_3 \cdot 32H_2O$ would simply be expressed in the cement notation as

$$C_6A\bar{S}_3H_{32}$$

The cement notation is used side-by-side with normal chemical formulae. The most important compounds present in Portland cement are

$$3CaO \cdot SiO_2 = C_3S$$
$$2CaO \cdot SiO_2 = C_2S$$
$$3CaO \cdot Al_2O_3 = C_3A$$

$$4CaO.Al_2O_3.F_2O_3 = C_4AF$$
$$\text{and } MgO = M$$

Normal Portland cement contains about 45% of C_3S, 27% of C_2S, 11% of C_3A and 8% C_4AF, with other materials accounting for the remaining 9%. Portland cement which is formulated to produce very little heat on setting has a higher percentage of C_3S, which accounts for 50% of the total, while the percentage of C_3S is restricted to 28% and that of C_3A to 4%. On the other hand, Portland cement which is designed for rapid development of strength is much richer in C_3S, which then accounts for up to 55% of the whole, while C_2S is restricted to 20% and less. The percentages of the other compounds are roughly the same as in normal Portland cement.

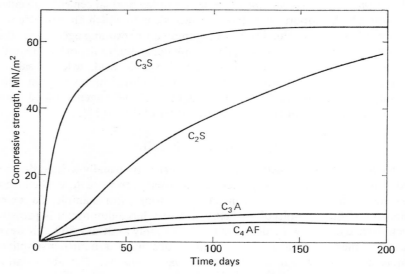

Fig. 1.1 Development of compressive strength of cement constituents

Figure 1.1 gives graphs of the hardening rates for pure samples of the main concrete constituents. It can be seen that the compressive strengths of C_3A and C_4AF are much lower than those of C_3S and C_2S, and also that C_3S hardens much more rapidly than C_2S. As hardening is inevitably accompanied by heat evolution, the percentage of C_3S is restricted in cements which must not give out too much heat, as for example when they are used in concrete mixes for casting thick sections.

THE APPLICATION OF THE PHASE RULE TO CEMENT CLINKER COMPOSITION

Binary relationships
In the study of cement clinker the following relationships are of importance:

The $CaO - SiO_2$ relationship.
The $CaO - Al_2O_3$ relationship.
The $Al_2O_3 - CaO$ relationship.

These are studied by plotting temperature versus composition on a graph. Figure 1.2 gives the relationship between CaO and SiO_2. As can be seen, the

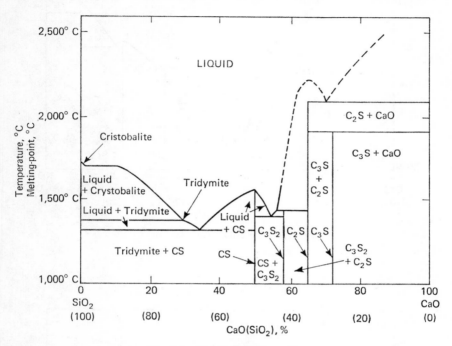

Fig. 1.2   The $CaO - SiO_2$ system

$CaO - SiO_2$ phase rule diagram consists of a number of eutectics side-by-side, caused by the intermediate compounds CS, C_3S_2, C_2S and C_3S. The most important compounds from the point of view of cement clinkers are C_2S, also known as calcium orthosilicate, which crystallizes in different forms depending upon the temperature, and C_3S, known as tricalcium silicate. Tricalcium silicate is normally unstable at ordinary temperatures but exists in a meta-stable equilibrium. Figure 1.3 gives the relationship between CaO and Al_2O_3.

The main compounds present in Portland cement are C_3A, $C_{12}A_7$ which used to be thought of as C_5A_3 (still quoted as this in some books), CA and CA_2.

CA is one of the main constituents of high alumina cement, which also contains $C_{12}A_7$.

Figure 1.4 gives the relationship between Al_2O_3 and SiO_2. The only stable intermediate compound between Al_2O_3 and SiO_2 is $3Al_2O_3.2SiO_2$ (A_3S_2),

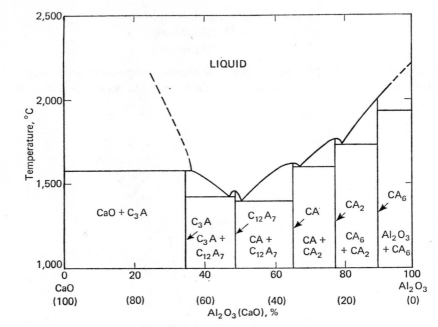

Fig. 1.3 The CaO–Al₂O₃ system

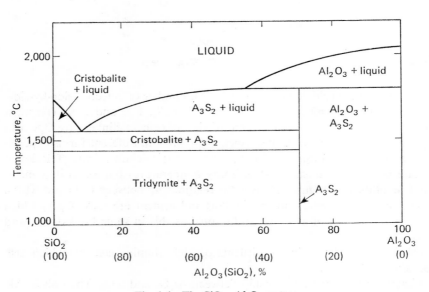

Fig. 1.4 The SiO₂–Al₂O₃ system

which is called mullite and is not generally present in Portland cement or other cements, although it is a constituent of refractories.

Triple relationships
In the study of cement systems one usually has to deal with material systems containing three or more primary constituents. To study a system with three constituents use is made of the Roozeboom diagram as shown in Fig. 1.5.

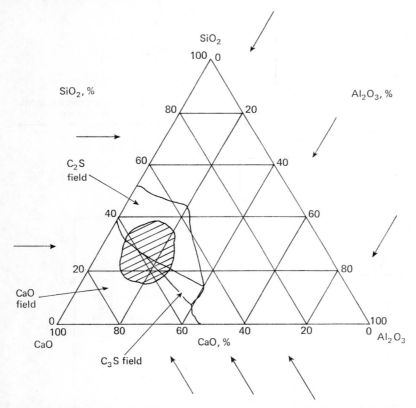

Fig. 1.5   The CaO–Al$_2$O$_3$–SiO$_2$ system: shaded area represents the Portland cement area

An equilateral triangle is drawn and a corner is assigned to each of the primary substances, in this case CaO, SiO$_2$ and Al$_2$O$_3$. The percentage of each is represented on a scale to the right of each point, the plane of the lines of equal composition being represented by the arrows drawn on the diagram. In this way the Roozeboom diagram can show the composition of any mix consisting of the three ingredients in question. On Fig. 1.5 the shaded portion which represents the Portland cement zone overlaps the C$_2$S, C$_3$S and C (CaO) fields. The centre of this area is equivalent to a melting-point of about 1,900°C,

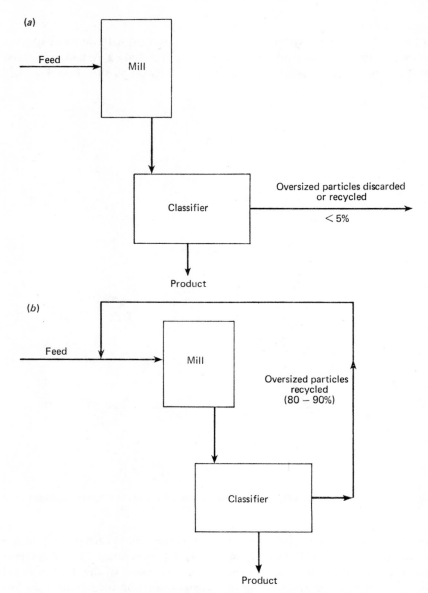

Fig. 1.6 Open-cycle grinding (*a*) and closed-cycle grinding (*b*)

which accords with melting-points of 2,570°C for pure CaO, 2,040°C for pure Al_2O_3 and 1,710°C for pure SiO_2. Similar triple diagrams can be drawn for $CaO-Al_2O_3-SiO_2$ and other systems.

THE MANUFACTURE OF CEMENT CLINKER

The composition of Portland cement is controlled by analysing the various constituents before blending and compounding them so as to produce a mix which accords with the specifications which apply to the given cement. One of the most important aspects is the relationship between basic constituents like CaO, and the various acidic constituents such as SiO_2, Al_2O_3 and Fe_2O_3.

British Standard Specifications (BSS) prescribe that the ratio

$$\frac{CaO}{2 \cdot 8SiO_2 + 1 \cdot 2Al_2O_3 + 0 \cdot 65Fe_2O_3}$$

must always be between 1·02 and 0·66, after due allowances have been made for the presence of other constituents not given in the equation. Equally, the ratio Al_2O_3/Fe_2O_3 must be more than 0·66.

For low-heat Portland cement, which is used in large castings such as dams, the maximum permitted lime content must not exceed

$$2 \cdot 4SiO_2 + 1 \cdot 2Al_2O_3 + 0 \cdot 65Fe_2O_3 \text{ by weight}$$

The percentage magnesia permitted is a maximum of 4% according to BSS, while the maximum percentage of gypsum allowed is 2·75%. Specifications in the United States and in Europe differ slightly from the British ones, but the over-all effect is the same.

The nature of compounds formed in the cement clinker is determined by both the composition of the feed and by the temperature to which the clinker is heated. The reason for this is that most of the compounds are present in various morphological modifications, and these structures are governed by the temperature range within which the compounds are formed. After cooling, the compounds tend to stay in their metastable equilibrium states.

After clinker has been formed and allowed to cool, it is ground. The normal form of grinding used today is closed-cycle grinding in which a small fraction of the clinker only is ground and the rest recycled (Fig. 1.6). As can be seen from Fig. 1.7, closed-cycle grinding tends to produce and remove a majority of particles of a size distribution just beneath the required limit, which means both a minimum of mechanical effort, as well as a more even distribution range. In open-cycle grinding, where the material is not recycled, or where the percentage recycled is very small, the vast majority of the particles are so-called 'fines', and the size distribution range is far more uneven.

The fineness of grinding is usually given in terms of the specific surface area of the cement. In Great Britain the surface area of normal Portland cement is specified to be not less than 2,250 cm²/g, while rapid hardening Portland cements should have surface areas not below 3,250 cm²/g.

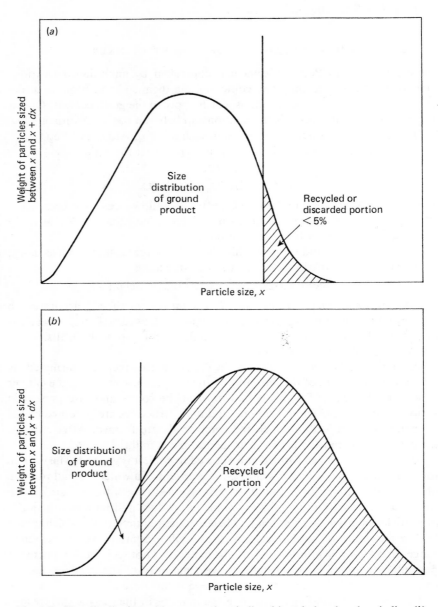

Fig. 1.7 Size distribution with open-cycle grinding (*a*) and closed-cycle grinding (*b*)

In general, fineness of grinding affects the quick hardening strength of concrete more than the long term strength.

The following table gives the percentage increase in compressive strength obtained, when a given formulation of cement is ground to 3,000 cm²/g instead of 2,000 cm²/g:

After	Increase in strength, %
7 days	18
28 days	6
90 days	4
365 days	1

It can be seen therefore that very fine and even grinding is particularly necessary with cement which is used when the concrete is expected to harden very quickly.

REACTION OF CEMENT WITH WATER

All the materials present in cement powder made from clinker are naturally anhydrous. When water is added to the cement, all are either attacked by water or dissolve in water to form hydrated compounds. The degree to which the compounds react with water depends both on the nature of the compounds and also on the fineness of grinding.

Typical reactions are the following:

$$C_3S + water \rightarrow aC\,a(OH)_2 + C_xS_y \text{ aqua.}$$
$$2C_3A + water \rightarrow C_2A \text{ aqua} + C_4A \text{ aqua.}$$
$$C_{12}A_7 + water \rightarrow 3CA \text{ aqua} + 4C_2A \text{ aqua} + Ca(OH)_2.$$
$$CA + water \rightarrow CA \text{ aqua.}$$
$$CA_2 + water \rightarrow CA \text{ aqua} + Al_2O_3.$$

C_2S is only very slowly attacked by water and forms a mere coating on the surface of the particles of hydrated silica.

When Portland cement is mixed with water, the individual constituents begin to react. C_3A reacts very quickly and C_4AF almost as rapidly. With many of the reactions $Ca(OH)_2$ is liberated which causes a rapid rise in the pH of the setting concrete. This is countered by adding SiO_2, in the form of sand, to the concrete mix so that any $Ca(OH)_2$ liberated reacts with it to form a series of hydrated calcium silicates.

The setting and hardening process taking place with Portland cement
When water is added to Portland cement, the cement grains are first acted upon to form a supersaturated solution from which a gel of crystals separates out. As this takes place, the viscosity of the concrete mix increases, and the mix achieves a good degree of plasticity. While this primary setting stage takes place a slight contraction occurs. The rate of primary setting is governed

largely by the quantity of gypsum ($CaSO_4.2H_2O$) and other calcium and magnesium salts present which slow down the rate of primary setting. If cement were formulated without the addition of gypsum, it would set immediately water was added and would therefore be quite useless for most purposes.

The secondary stage is the formation of a rigid gel of crystals with colloidal dimensions and also, simultaneously, larger crystalline products. The crystals formed are of calcium hydroxide, hydrated calcium aluminate and complex sulphoaluminates. During this secondary process there is a slight expansion of the concrete mix. Usually the primary and the secondary processes occur virtually simultaneously. The increase in internal pressure within the gel structure produces the increasing compressive strength of the concrete.

The final process takes place very much more slowly. It is the chemical interaction with carbon dioxide from the atmosphere, so that in the end the concrete consists of a mass of coarse crystals of calcium carbonate, hydrated silica, ferric oxide and alumina.

The role of water in the setting of cement
The water which is part of the setting cement mass can be considered to fulfil four different functions:

1 *Free water* is not held to the concrete by any special forces and largely evaporates during the setting of the concrete.

2 *Capillary water* is held by capillary forces within the pores of the concrete and does not readily evaporate.

3 *Gel water* has the purpose of forming the gel which constitutes the primary and secondary phases in the setting of the concrete. During the final hardening process it is changed into combined water.

4 *Combined water* is an integral part of the concrete structure.

The following list gives the density of a number of typical fully hydrated compounds which are present in most concretes:

$3CaO.2SiO_2.3H_2O$ ($C_3S_2H_3$) (tobermorite)	2.44 kg/dm³
$3CaO.Al_2O_3.6H_2O$ (C_3AH_6)	2.52 kg/dm³
$3CaO.Al_2O_3.8H_2O$ (C_3AH_8)	2.13 kg/dm³
$4CaO.Al_2O_3.12H_2O$ (C_4AH_{12})	2.15 kg/dm³
$2CaO.Al_2O_3.8H_2O$ (C_2AH_8)	1.95 kg/dm³
$3CaO.Al_2O_3.3CaSO_4.32H_2O$ ($C_6A\bar{S}_3H_{32}$)	1.73 kg/dm₃
$3CaO.Al_2O_3.CaSO_4.12H_2O$ ($C_4A\bar{S}H_{12}$)	1.95 kg/dm³

Altogether, fully set cement has as much as 20—25% of its weight in the form of combined water. The density of hydrated cement is a good deal lower than that of anhydrous cement. For example, anhydrous Portland cement has a density varying between 3·0—3·2 kg/dm³, while the density of the set concrete (part due to cement only) is only 2·15 kg/dm³.

The effect of adding gypsum to concrete mixes

To prevent over-rapid initial setting of the concrete mix certain retarders are always added. The most common of these is gypsum. It retards the primary set of cement because it reacts with calcium aluminates to form two practically insoluble compounds, namely $3CaO.Al_2O_3.3CaSO_4.32H_2O$ and $3CaO.Al_2O_3.CaSO_4.12H_2O$. Setting is also slowed down when the Portland cement is mixed with excess of $Ca(OH)_2$. In fact, mixtures of $CaSO_4$ and $Ca(OH)_2$ cause concrete to set more slowly than when $CaSO_4$ is used on its own. Other retarders are calcium chloride, calcium sulphate, calcium iodide and sodium carbonate–sodium silicate solutions. Lead, copper and zinc salts also serve to retard the setting of concrete, while substances such as sugar and borax can seriously retard the development of strength, though they have little effect on the primary and setting processes.

Heat liberation during concrete setting

When concrete sets, heat is always liberated because the reactions of the various cement constituents with water to form hydrates are markedly exothermic.

The heats of hydration of the different compounds differ quite strongly, however, as does the percentage of this heat which is evolved during the initial stages of setting. The following table gives the heat of hydration of a number of compounds commonly present in concrete and also the percentage of heat liberated within 48 hours of mixing the concrete.

Compound	Heat liberated, J/g	Heat liberated within 48 hr, %
$3CaO.Al_2O_3$ (C_3A)	864	73
$3CaO.SiO_2$ (C_3S)	500	83
$4CaO.Al_2O_3.Fe_2O_3$ (C_4AF)	418	40
$2CaO.SiO_2$ (C_2S)	258	16

The heat of hydration of pure CaO is 1,160 J/g and that of pure MgO is 840 J/g.

The effective heat of hydration of a given Portland cement mix is thus determined by the individual heats of hydration of its components. The amount of heat given off by a Portland cement at the initial stages of setting is also governed by the degree of fineness of grinding.

With a normal type of Portland cement the total heat of hydration involved is of the order of 490 J/g, while rapid-hardening Portland cement evolves 500 J/g and the low-heat type only evolves 400 J/g. But whereas in normal Portland cement 52% of the heat, or 256 J/g, is evolved within three days of setting, 62% of the total heat of setting, or 310 J/g, is evolved within three days in the case of rapid-hardening Portland cement, and 48%, or 192 J/g, is evolved within three days from the low-heat type of cement.

Temperature rises in massive structures due to the large quantities of heat evolved during setting of concrete are considerable, even when low-heat cement is used. Internal temperatures of large sections may rise by as much as 40°C. Attempts have been made in several contracts to counteract this appreciable temperature rise by mixing the concrete with ground ice. The latent heat of the ice, which amounts to 320 J/g, goes some way to absorb some of the heat of setting of the concrete. This method was recently used during the construction of the 2-m thick and 18-m diameter foundation slabs for the Marina City contract in Chicago, which had to support circular apartment blocks of 16 floors of parking ramps with 40 floors of dwellings above, a total height of 180 m above street level.

CHEMICAL RESISTANCE OF PORTLAND CEMENT

The various compounds which are present in Portland cement are readily attacked by water, as well as many salt solutions and acids.

1   *Water*. When concrete is in permanent contact with water, the calcium salts present tend to hydrolyse and leach out, leaving residues of silica, iron oxides and alumina. However, such reactions tend to be concentrated on the surfaces of the concrete only, leaving the body of the concrete unaffected unless the concrete is particularly porous for some reason. When the water is acidic, the rate of attack is very much increased. Acidic waters often occur in nature due to either the presence of natural humic acids, or of dissolved carbon dioxide. In industrial areas, waters frequently contain free sulphuric acid, in addition. While neutral water has a pH of 7·0, water which has CO_2 dissolved in it usually has a pH of about 5·7, humic acid solutions have a pH of between 3·5 and 4, while water from industrial areas can have pH values as low as 2·5.

When acidic water acts upon concrete for long periods of time the surface becomes very soft and starts to disintegrate.

2   *All acids* attack concrete. The strong mineral acids have the most rapid effect, particularly if soluble calcium salts are formed. Organic acids such as lactic acid, acetic acid or citric acid also tend to soften concrete. An interesting type of corrosion of concrete takes place when coal or coke is stored against concrete walls. The sulphur contained in the fuel oxidizes to sulphur dioxide which forms sulphurous acid with water, causing the concrete to be attacked readily.

3   *Alkalis* on the other hand do not generally have any adverse effect upon concrete.

4 *Sulphate solutions* such as sodium, potassium and other sulphates very readily attack concrete. A typical reaction which takes place is the following:

$$Ca(OH)_2 + Na_2SO_4 + 2H_2O \rightarrow CaSO_4 . 2H_2O + 2NaOH$$

The calcium sulphate is leached out, as is the sodium hydroxide, thus weakening the structures. Hydrated calcium silicates are not, however, markedly affected by dissolved sodium and potassium sulphates but are attacked by magnesium sulphate solutions:

$$3CaO . 2SiO_2 \, aqua. + 3MgSO_4 \rightarrow 3CaSO_4 . 2H_2O + 3Mg(OH)_2 + 2SiO_2 \, aqua.$$

The resistance of concrete to sulphate solutions is very much increased by methods of steam curing, particularly under superheated steam conditions. The rate of attack by sulphates is very markedly dependent upon the temperature. For example, concrete is often corroded by magnesium sulphate contained in sea water in subtropical and tropical regions. In temperate areas, on the other hand, the rate of attack on concrete by sea water is not excessive.

5 *Chlorides.* Sodium chloride has, in general, little effect on concrete but the effect of ammonium chloride is almost the same as that of an acid. Calcium chloride, unless present in large concentrations, has little effect.

6 *Organic materials.* Certain organic materials damage concrete badly. Glycerine has a strong corrosive effect as have many drying oils such as linseed oil, etc., which reacts with free calcium oxide. Mineral oils, when not adulterated with acids, have no effect upon concrete at all.

HARDENING OF CONCRETE AND DEVELOPMENT OF STRENGTH

The strength of concrete is dependent upon the following factors.

Composition
As already mentioned, the strongest constituent in the clinker mix is C_3S, while most other compounds used have far lower compressive strengths. The component C_2S, which sets very slowly indeed will harden finally to a strength not far below that of C_3S, but compounds like C_3A, C_4AF and others harden to very much weaker products.

Water content and air content
Two general laws have been formulated to give the relationship between water or air content, and final compressive strength.

The Abrams equation deals with concretes where the percentage of air voids is below 1%. It is commonly given as

$$S = AB^{-x}$$

where x is the water/cement ratio by weight, S the strength and A and B constants depending upon the nature of the concrete, prevailing temperature, etc.

The Feret equation deals mainly with air entrained concretes and is given as

$$S = K\left(\frac{c}{c+w+a}\right)$$

where K is a constant, c the volume of cement in the mix, w the volume of water in the mix and a the volume of entrained air in the mix.

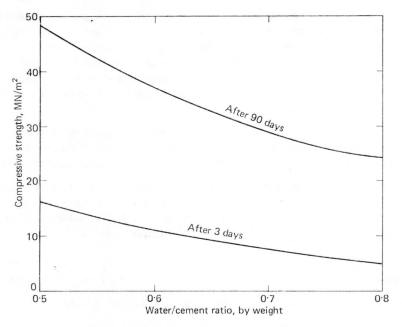

Fig. 1.8 Compressive strength versus water/cement ratio

Figure 1.8 shows how the compressive strength of a concrete made from 1 part cement, 2 parts sand and 4 parts ballast after 3 and 90 days is affected by the water/cement ratio used.

Contact with water during the curing (final setting) period
When concrete is cured in the presence of water, as for example when it is submerged or surrounded by damp materials, a better development of strength takes place. Concrete cured under water usually has between 5 and 10% higher compressive strength after 90 days than concrete which has been kept in air for the same period of time.

Temperature

When concrete is allowed to harden at higher ambient temperatures, the development of strength takes place far quicker, even though the final strength achieved will probably be the same. Figure 1.9 shows the relative compressive strengths when Portland cement is allowed to set at different temperatures after 3 days and after 28 days. The phenomenon of more rapid hardening is

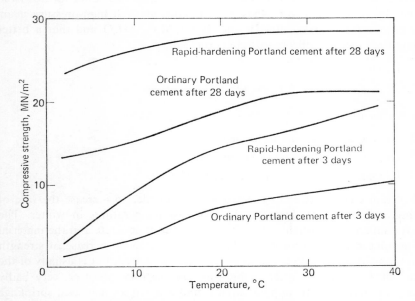

Fig. 1.9   Effect of ambient temperature on the strength of Portland cement

accentuated when rapid-hardening Portland cement is used. After 3 days' setting at a temperature just above the freezing-point, rapid-hardening concrete achieves a compressive strength of 3 MN/m² as against a compressive strength of 18 MN/m² after 3 days at 35°C.

STEAM CURING OF CONCRETE

When concrete made from Portland cement is cured at temperatures of 100°C and above in steam, at first a slight contraction takes place, followed by an expansion. In the same way, the compressive strength first goes down and then up. Such steam curing does not cause the concrete to achieve its final strength particularly rapidly, but it does produce a rapid development of initial hardness. For example, 16 hr curing in steam at 100°C on a typical concrete with cement/sand/ballast ratio of 1:2:4 produces the same hardness as 3 days' air curing in moist air at 18°C.

C

High-pressure-steam curing, on the other hand, often achieves higher final compressive strengths than prolonged air curing.

If concrete is cured for periods of about 48 hr under the action of steam at 15 bar (gauge), the compressive strength of the material can be as much as 30 % higher than that of similar concrete products which have been cured in moist air for 28 days. High-pressure-cured concrete products have considerably lower drying shrinkages than normal concretes and also suffer far less from attack by sulphates. The reasons given for this are that high-pressure-steam curing eliminates free $Ca(OH)_2$ and $3CaO.Al_2O_3.6H_2O$ and that a better crystallized calcium silicate hydrate is formed.

CONCRETE ADDITIVES

Chemicals are added to concrete for three main reasons:

1 To accelerate the rate of hardening.

2 To slow down the rate of hardening.

3 To make the concrete more waterproof.

Calcium chloride $(CaCl_2)$ is often added in order to increase the rate of strength development during periods of low temperatures in winter. The maximum added is usually about 2 % by weight of concrete and the material is introduced in the form of a solution. The degree of increase of strength development varies with the kind of concrete concerned, but is usually of the order of 3—7 MN/m^2 after 24 hr. Calcium chloride, however, very badly affects the resistance to sulphate attack, and also causes increased shrinkage during drying out. Other salts such as sodium carbonate, sodium silicate, aluminium chloride, etc., also increase the rate of strength development somewhat, but without the same deleterious effect upon sulphate resistance.

When it is desired to retard the hardening of concrete for some reason or other, such as the production of exposed aggregate, sugar solutions or dextrin solutions are employed. These solutions are usually applied to surface layers of concrete before it has started to set, and markedly retard the hardening of the surface. The concrete can then be brushed away with a stiff brush, and in this way the aggregate is exposed.

A large number of different chemicals are added to concrete mixes as so-called 'integral waterproofers'. These include calcium, sodium, potassium and aluminium soaps of fatty acids, sodium silicate, sodium silicofluoride, glues, mineral oils, jellies, waxes, proteins, celluloses and various kinds of vegetable oils and resins.

The effect of these additives is variable. In some cases they tend to discourage pore formation, and in others they have the property of increasing the angle of capillary contact within the pore structure of the concrete, thus reducing capillary attraction to water. Integral waterproofers of the type mentioned should never be used with high alumina cements, as the strength is very badly affected by them.

SURFACE WATER PROOFING OF CONCRETE

Two types of materials are commonly used:

1 Materials of an organic origin.
2 Materials of an inorganic origin.

1. The most common material of all is a solution of bitumen in a solvent such as benzene, which is painted onto the surface. Bitumen is also applied hot by spray or brush, or in the form of an emulsion. Chlorinated rubber paints, synthetic resin lacquers and wax coatings are frequently brushed or sprayed on, while rubber latex is another extremely popular material. Drying oils may be brushed on and react with the concrete surface to form impervious layers of calcium soaps as the esters are saponified by free calcium hydroxide in the concrete.

$$
\begin{array}{l}
C_nH_mCOOCH_2 \\
\quad | \\
C_nH_mCOOCH + 3Ca(OH)_2 \rightarrow 3(C_nH_mCOO)_2Ca + 2 \\
\quad | \qquad\qquad\text{(lime)} \qquad\qquad\qquad \text{(calcium soap)} \\
C_nH_mCOOCH_2 \\
\quad\text{(drying oil)}
\end{array}
\qquad
\begin{array}{c}
H \\
| \\
HO-C-H \\
| \\
HO-C-H \\
| \\
HO-C-H \\
| \\
H \\
\text{(glycerol)}
\end{array}
$$

These calcium soaps are extremely effective as water stops. Impregnation of surfaces with mineral oil solutions is of a less permanent character.

Finally, films of polyethylene and PVC are frequently glued on to concrete surfaces as protection.

2. Surfaces are often painted with sodium silicate, which hydrolyses on the surface of the concrete as follows:

$$Na_2SiO_3 + H_2O \rightarrow 2NaOH + SiO_2$$

The silicon dioxide reacts with free lime to form a firm and water-repellant crystalline structure, while the sodium hydroxide is usually simply leached away. Magnesium and zinc silicofluoride solutions are painted on, which form complex silicofluoride structures with lime at the surface of the concrete:

$$Zn\ SiF_6 + Ca(OH)_2 \rightarrow Ca\ SiF_6 + ZnO + H_2O$$

A special method which is commonly applied to make precast concrete objects waterproof is to expose them to silicon tetrafluoride (SiF_4) vapour. A firm and impermeable coating of calcium silicofluoride is produced at the surface and in the surface capillaries of the concrete objects:

$$2SiF_4 + 2Ca(OH)_2 \rightarrow Ca\ SiF_6 + CaF_2 + SiO_2 + 2H_2O$$

1.2 Non-Portland cements

HIGH ALUMINA CEMENT (CIMENT FONDU)

High alumina cement is made from a mixture of limestone and bauxite, a material which contains roughly 55% Al_2O_3 and 24% Fe_2O_3 with smaller quantities of SiO_2, TiO_2 and about 15% combined water. The two materials are heated together to about 1,600°C to produce a dark clinker. High alumina cement is produced by grinding the material to a specific surface area of 2,300—2,900 cm²/g and consists of the following compounds:

$$C_{12}A_7,\ CA,\ C_6A_2F,\ CF\ and\ C_2F$$

No additions are made during the grinding of the cement as in the case of Portland cement. The rate of setting is primarily governed by the relative abundance of $C_{12}A_7$, which sets very quickly and of CA, which only sets slowly.

When water is added to high alumina cement, a number of complex hydrates such as C_2AH_8 and CAH_{10} are formed, which produce a final gel of C_3AH_{10} with hydrated aluminium oxide. The iron is present in the hydrated form as C_3FH_6 and also as ferric hydroxide.

High alumina cement needs far more water during setting than Portland cement. In the final form it usually contains as much as 35—50% combined water. It swells a good deal more than Portland cement does, during the hardening process, but it also contracts more during the initial setting process.

High alumina concrete mixes use a slightly higher percentage of sand than equivalent Portland cement mixes and need a minimum water/cement ratio of 0·4. The setting time of high alumina concrete mixes is roughly the same as that of Portland cement mixes, but the final strength developed is a good deal higher. Compressive strengths of 1 part cement:3 parts of sand concretes after 7 days of 90 MN/m² and for 1 part cement:2 parts of sand:4 parts of ballast of 60 MN/m² after 7 days have been quoted. These are much higher figures than those achieved with equivalent Portland cement mixes, which show barely half these compressive strengths after this time.

High alumina cement, furthermore, develops its strength very quickly, and often achieves compressive strengths up to 70—80 MN/m² at the end of a single day's air curing.

Unlike Portland cement mixes, high alumina cement mixes are very adversely affected by the presence of calcium chloride in the mixing water, and sea water must never be used for mixing alumina cements. Surfaces of alumina cement cannot be waterproofed by sodium silicate or siliconate, but instead drying oils or magnesium silicofluorides should be used.

High alumina cement is sometimes mixed with Portland cement. This practice reduces the setting time and also the final strength of the mixes in comparison with individual cements.

Chemical resistance

High alumina cement is not as readily attacked by water containing sulphates, the reason for this being the absence of free calcium hydroxide in the set concrete. High alumina cement has also a much better resistance to acid waters than Portland cement. On the other hand, high alumina concrete reacts readily with alkalies or even alkali carbonates:

$$K_2CO_3 + CaO . Al_2O_3 + H_2O \rightarrow CaCO_3 + 2KOH + Al_2O_3$$
$$2KOH + Al_2O_3 \rightarrow 2K(AlO_2) + H_2O$$

The soluble potassium aluminate is then leached out.

High alumina concrete is more stable to high temperatures than Portland cement concrete and is often used for the bonding of refractory bricks in furnaces and other equipment. Alumina concretes are usually stable up to 1,350°C and if additions of carborundum, chrome, etc., are made, stability to temperatures up to 1,600°C can be achieved.

Pozzolana cements

Natural pozzolanas are materials which are able to combine with lime to form cements. Most of the natural pozzolanas are of volcanic origin and are found, naturally, in countries such as Italy, Japan, New Zealand and others where a good deal of volcanic activity takes place. Pozzolanas can also be made artificially by burning various clays and shales. Even the fly-ash obtained from coal-fired power stations is technically known as an artificial pozzolana. As is to be expected, the composition of both natural and artificial pozzolanas varies considerably. It is usual to mix 1 part of slaked lime putty by volume with between 2 and 3·5 parts of ground pozzolana and about 8—12 parts of aggregate.

Pozzolana concretes usually set by reactions such as

$$2(Al_2O_3 . 2SiO_2) + 7Ca(OH)_2 + water \rightarrow 3CaO . 2SiO_2 \text{ aqua.}$$
$$+ 2(2CaO . Al_2O_3 . SiO_2) \text{ aqua.}$$

The final strength of pozzolana concrete varies with the amount of lime present. For most pozzolanas the optimum percentage of lime for a maximum compressive strength is around 26—28%, although this varies with the composition of the pozzolana itself. The compressive strength of pozzolana concrete is not particularly high, seldom exceeding about 16 MN/m², even several years later on. The main advantages of pozzolana cement concrete over other forms are their excellent resistance to sea water and many other solutions, particularly sulphates, and their relative cheapness.

Blast furnace cement

In blast furnaces the quantity of slag obtained usually amounts to one ton per ton of iron, and in consequence it is advantageous to find uses for the material. Blast furnace slag usually consists of 40—50% CaO, 35% SiO_2, between 10 and 20% Al_2O_3 with varying smaller quantities of MgO, FeO, MnO_2 and up to 2% of sulphur combined with various metals. Granulated slag on its own has virtually no cement properties, but it can be mixed with Portland cement, which appears to activate it to perform like a cement itself.

Another method of making use of blast furnace slag to produce a cement is to mix it with up to 30% of lime, using an addition of 1% sodium sulphate to accelerate the setting process. Slag cement is used for underground foundation work and also for the construction of structures which are likely to be exposed to sea water, because of the good resistance of such cement to sulphates. The material hardens very slowly and has a compressive strength below that of Portland cement even after long periods of time.

For this reason it is more usual to mix blast furnace cement with Portland cement rather than use it on its own.

Magnesium oxychloride or Sorel cement

Anhydrous magnesium chloride and magnesium oxide are mixed together with water in the ratio by weight of 1·5:0·8:1·8. Sorel cement is a very rapid hardening cement and its final strength may be as high as 55 MN/m^2. The chemical formula for hardened Sorel cement is a macromolecular form which approximates to the composition $3MgO.MgCl_2.11H_2O$, in which some of the material is present as the oxychloride, $Mg(OCl)_2$. The material is very readily attacked by water which tends to leach out the magnesium chloride. It is commonly used in conjunction with an inert filler and various pigments as a flooring material. The surface should always be painted with a water repellent material.

Sorel cement has a corrosive action on steel objects which are embedded in it.

A cement which is somewhat similar to Sorel cement is magnesium oxy-sulphate cement, which is widely used as a binder for wood wool slabs.

Literature Sources and Suggested Further Reading

1. BANKS, R. F., and KENNEDY, H. L., *The Technology of Cement and Concrete*, 2 volumes, Wiley, New York (1955)

2. BOGUE, R. H., *The Chemistry of Portland Cement*, Reinhold, New York (1955)

3. BOYNTON, R. S., *Chemistry and Technology of Lime and Limestone*, Interscience, New York (1966)

4. CEMENT AND CONCRETE ASSOCIATION, Technical Information and Brochures (1958)
5. CHILDE, H. L., *Concrete Products and Cast Stone*, Cement and Concrete Association (1961)
6. HELLSTRÖM, B. O., GRANHOLM, H., and WÄSTLUND, G., *Betong*, 2 volumes, Natur och Kultur, Stockholm (1958)
7. LEA, F. M., *Chemistry of Cement and Concrete*, Arnold, London (1956)
8. ROBSON, T. D., *High Alumina Cements and Concretes*, Wiley, New York (1962)
9. TAYLOR, H. F. W., *The Chemistry of Cements*, 2 volumes, Academic Press, New York (1964)

Address: The Cement and Concrete Association, 52 Grosvenor Gardens, London SW1

Chapter Two   Special Concrete Products, Gypsum and Asbestos

2.1 Coloured concrete

WHITE

Normal Portland cement is grey in colour due to the presence of iron oxides. In white concretes these must be kept down absolutely and therefore the raw materials for white cement are usually chalk and china clay. Oil fuels are used for burning, to prevent adulteration with coal ash, which also often contains iron oxides. White cement has a strength which is usually lower than that of standard Portland cement.

Typical analysis figures for white cements are:

CaO	64%	Al_2O_3	6%
SiO_2	24%	Fe_2O_3	$< 0.8\%$

The rest is made up of bound water, magnesia, sulphates, etc.

BLACK

Black concretes are difficult to formulate. The most usual additions made are manganese dioxide and carbon black. The former tends to give a rather greyish cast, and cannot be used for making a concrete which is truly jet black. Carbon black gives a deeper colour, but the strength of the concrete is very badly affected.

OTHER COLOURS

Reds are produced by the admixture of iron oxides, which are also used for obtaining certain yellowish and brownish types of concrete. The reds are obtained by the use of Fe_2O_3, so-called red oxide of iron, the yellows by the use of ochre, which is a complex double compound of red oxide of iron and ferric hydroxide, while brown grades are produced by the use of umbers, which are double compounds of Fe_2O_3 and MnO_2.

Green colours are obtained by the use of chromium oxide, while blue colours are produced by using ultramarine blue and cobalt blue. Ultramarine blue tends to fade as it combines with free $Ca(OH)_2$, but this reaction is somewhat counteracted by atmospheric acids. Cobalt salts are more permanent than ultramarine, but also far more expensive.

The pigment content is usually of the order of 5—10% by weight of the cement used.

Other methods of producing coloured concrete are to paint the surface with an aniline dye solution when semi-set or to use coloured ballast. This is then exposed by using a dextrin solution to retard the setting of the cement, so that it may easily be brushed or washed away.

2.2 Concrete emulsions

Cement is sometimes mixed with synthetic resin solutions or rubber latex to give a useful flooring finish. Rubber latex is first stabilized and then mixed with either high alumina cement or Portland cement together with various organic fillers such as sawdust, as well as pigment additions. These are used in layers about 0·5 cm thick as flooring finishes and also as jointing compounds. Where floors have to be constructed which are resistant to oils and greases, finishes are compounded using high alumina cement together with neoprene latex, or either PVC or PVAC emulsions. Other materials that can be used are bitumen/cement emulsions.

STABILIZATION OF SOIL

Cement grout can be injected into loose soil in order to get sufficient consolidation, to enable it to act as a load-bearing foundation. The method employed is to use solutions of calcium chloride and sodium silicate which react together as follows:

$$2Na_2SiO_3 + CaCl_2 + aqua \rightarrow CaO.2SiO_2 \ aqua + 2NaCl + 2NaOH$$

This process is carried out by using separate perforated pipes for each solution. These are driven into the soil, followed by pressure injection of the liquids.

CONCRETE FOR RADIATION SHIELDING

Concrete that is employed for such purposes as biological shields in reactor technology, for shielding X-ray equipment and other 'hot' areas is formulated by using a high proportion of barium sulphate in the mix. This is added in the form of barytes sand or as barytes aggregates, using as much as 80% of $BaSO_4$ in the cement mix. Such concrete has a density of 3·3 kg/dm³ and a strength of about 14 MN/m² after seven days. Other materials which are used in concrete employed for radiation shielding are quantities of magnetite (Fe_3O_4) and limonite ($2Fe_2O_3.3H_2O$). Such aggregates serve to produce concretes which have densities in excess of 4·5 kg/dm³. In general, the resistance of a concrete shield to radiation is proportional to the density of the concrete used.

2.3 Gypsum plasters

Gypsum is a natural mineral which has the formula $CaSO_4.2H_2O$, a hardness on the Moh scale of only 2 and a density of $2 \cdot 317$ kg/dm³. When gypsum is heated carefully, it loses 75% of its water of constitution and forms the hemihydrate $CaSO.\frac{1}{2}H_2O$. This material is then known as 'Plaster of Paris'. The heating is carried out at around 150°C in shallow open iron pans. Further heating produces anhydrite, which is simply anhydrous $CaSO_4$, while calcining up to 1,100—1,200°C causes the dissociation of part of the gypsum into CaO and SO_3. All three types of materials are used for the production of plasters.

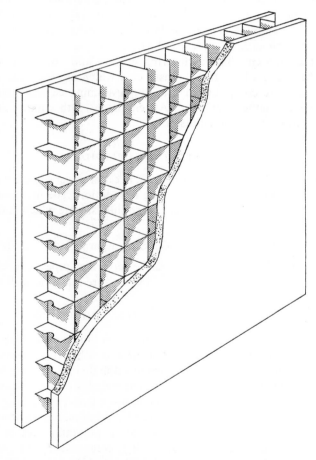

Fig. 2.1   Cellular partition using gypsum (By courtesy of ICI Limited)

When the hemihydrate is mixed with water, almost immediate setting takes place, to reform the gypsum. This takes place accompanied by the evolution of heat and expansion of the material. To reduce the speed of setting it is necessary to add $0\cdot1\%$ of keratin, which acts as a retarder.

The anhydride is often referred to as Parian plaster and is much slower setting than the hemihydrate, especially if it contains an appreciable portion of free CaO. In such cases it is necessary to add accelerators in the form of potassium, aluminium, ferrous or zinc sulphates, to the extent of about $0\cdot5\%$ by weight. Gypsum is widely used for the manufacture of a large range of acoustic tiles, boarding, impregnated wood wool slabs and similar materials. These are employed as fire-resistant panels, for reduction of noise in buildings, dry-wall panels and ceiling boarding. Figure 2.1 shows a cellular partition using gypsum.

Gypsum plasters are usually mixtures of the calcium sulphate hemihydrate with perlite, which is a natural glass of a granite composition containing about 75% SiO_2 and 13% Al_2O_3, which has been caused to have a very porous structure during its formation.

Vermiculite, which will be dealt with later on, is also used in admixture with gypsum hemihydrate, either alone or together with perlite.

2.4 Lightweight concretes

The term *lightweight concrete* covers three main groups of materials, namely:

1 No-fines concrete.
2 Lightweight aggregate concrete.
3 Aerated or gas concrete.

NO-FINES CONCRETE

This term is given to concrete composed solely of a mixture of cement and coarse aggregates having a size range between 1 and 2 cm in diameter. In addition, the material has a large number of voids in the structure, which gives it a much lower density than normal concrete.

No-fines concrete is usually prepared by employing a cement/aggregate mix in the ratio of $1:8$ by weight with a water/cement ratio of $0\cdot40$.

When no-fines concrete sets, the rate of shrinking is higher than with normal aggregate concrete but the degree of final shrinkage usually amounts to only about 60% of that normal. The degree of drying shrinkage depends upon the kind of aggregate used. Similarly, due to the fact that no-fines concrete has a cellular pore structure rather than continuous capillaries as in the case of normal concrete, water penetration is very low so that it is permissible to use solid walls of the material for external faces. Finally, the surface of no-fines concrete has excellent sound absorbing qualities. While the absorption coefficient for sound in the frequency range 125 Hz to 2 kHz equals $0\cdot02$

for both brickwork and normal concrete, the absorption coefficient of acoustic no-fines concrete walls, 18 cm thick and made with 5/10 mm aggregate is 0·55, whereas for acoustic no-fines concrete walls with aggregate size between 10—20 mm the absorption coefficient is around 0·7.

The general properties of 1:8 no-fines concrete are given in Table 2.1.

TABLE 2.1

Aggregate	Compressive strength after 28 days MN/m²	Drying shrinkage at 50°C and 17% relative humidity, %	k-value (thermal conductivity), W/m.degK
Rounded gravel	8·6	0·018	0·49
Crushed limestone	6·8	0·016	0·49
Crushed granite	7·5	0·018	0·47
Blast furnace slag	4·8	0·025	0·38

The density of no-fines concrete varies between 67 and 75% of that of normal concrete using the same aggregate. Its main advantage during building is that even if the no-fines concrete is cast to considerable heights, there is a very low hydrostatic pressure at the bottom of the mould so that it is possible to cast considerable heights of concrete using only simple plywood shuttering.

LIGHTWEIGHT AGGREGATES

Lightweight aggregate concrete uses furnace residues which are held together by normal Portland cement, to produce lightweight blocks and other sections of concrete. The material is now usually called 'furnace clinker' and no longer 'breeze' to avoid confusion with the name given to small particles of coke. Furnace clinker has an extremely variable composition, but consists mainly of SiO_2 and Al_2O_3 with up to 30% Fe_2O_3, up to 10% CaO and up to 5% MgO. There are also alkali metals present in the form of ions, and the percentage of sulphur, given as SO_3, can amount to any value between 1 and 12%.

The clinker is usually obtained from chaingrate stokers and contains un-burnt carbonaceous matter. However, it is desirable that the percentage combustible matter contained in the clinker should be as low as possible, as the compressive strength of the concrete is very badly affected by it. In addition a high percentage of combustible matter increases the drying shrink-age of the concrete. The percentage expansion under damp conditions is virtually doubled if a clinker with 30% of combustible matter is substituted for one possessing only 5% of combustible matter. For these reasons the content of carbonaceous matter in furnace clinker to be used for the pro-duction of lightweight aggregate blocks is rigidly specified in most countries.

BSS 1165:1957, Appendix C, specifies the following limits with regard to the clinkers to be used in lightweight aggregate concretes.

1 Aggregates for use in plain concrete for general purposes: combustible matter less than 10%.

2 Aggregates for use in *in situ* concrete for interior work not exposed to damp conditions: combustible matter less than 20%.

3 Aggregates for precast clinker concrete blocks: combustible matter less than 25%.

Certain coals have also been found to be particularly harmful when present in clinker to be used for aggregate concrete, because they cause progressive expansion of the concrete which produces fissuring. The reason for this is that the coal oxidizes spontaneously. The trouble is found mainly with coal which has intermediate properties between low-ranking brown coal and high-carbon-content coals such as anthracite. Clinker concrete is also liable to cracking if particles of hard burnt lime, magnesia or calcium sulphate are present. These react with water, producing rapid expansion. The phenomenon can be prevented by leaving the clinker exposed to the weather for a few months prior to use.

If the percentage of sulphur in the clinker does not exceed 1%, little trouble is likely to be experienced on this account. Sulphur compounds, however, accelerate the rate of corrosion of embedded reinforcement rods and mesh.

The use of foamed blast furnace slag

Blast furnace slag has about 40% free CaO, 30% SiO_2 and 15% Al_2O_3 with varying quantities of other constituents mainly in the form of magnesia and as sulphates. The iron oxide seldom exceeds 1%.

The slag comes from the blast furnace at about 1,500°C and when it is allowed to cool slowly, produces a hard material which is very useful as aggregate for heavy concrete and for foundation work. If the molten slag is quenched with water, steam is trapped inside the mass which expands it to produce foamed slag.

Foamed slag has bulk densities varying between 0·32 to 0·48 kg/dm³ for grades above 1 cm in diameter, although it may go as high as 0·8 kg/dm³ for some of the foamed slag fines.

Foamed slag has a slightly lower sulphur content than the original slag from which it is produced, and reacts well with cement to produce aggregates for concrete building blocks and also for use as an insulating roof screed. The strength of foamed slag concrete varies with the degree of compaction employed and also the cement/slag ratio. A fully compacted mix with a 1:4 cement/aggregate ratio has a compressive strength after 28 days of 22 MN/m². The strength of a 1:10 mix used in the semi-dry form is, however, only 3·4 MN/m² after 28 days.

Expanded shale, clay, pulverized fuel ash, etc.

These materials are known by a variety of trade names such as 'Leca', 'Aglite' and 'Haydite'.

They are made by heating small pellets of the material very quickly up to a temperature of 1,350°C. As this is done, gases are produced within the body of the pellets which expand them to a volume some 600—700% of their original value.

Concretes made from expanded clay have drying shrinkages varying between 0·04 and 0·07 and thermal conductivities between 0·216 and 0·575 W/m.degK. Figure 2.2 shows how the density of the concrete is related to the cube compressive strength after 28 days.

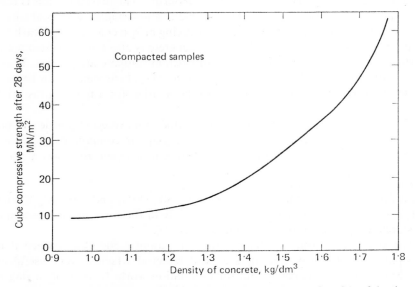

Fig. 2.2 Compressive strength of expanded clay concrete as a function of density

Pulverized fuel ash production in the United Kingdom amounts to some 6 million tons per annum, and in consequence there is a considerable need to find as many useful outlets for the material as possible. Expanded pulverized fuel ash is marketed under the trade names 'Terlite' and 'Lytag', and is made by making pellets of the ash, using a fine spray of water which impinges upon tilted rotating pans filled with the dry finely powdered material. Afterwards, the pellets are heated rapidly to 1,200°C. Because the flyash contains a fair proportion of unburnt coal, very little additional fuel is needed for the roasting process.

Concrete which is made from sintered pulverized fuel ash has a drying shrinkage between 0·04 and 0·06%, and a moisture expansion of the same order. The compressive strength varies with the density and is given in Fig. 2.3.

Slate is another material which is used as a lightweight aggregate after expansion. When heated strongly the material expands to give a product which is very light indeed and can float on water. Concretes can be made using the material in 1:10 ratio of cement to slate. These have a density as low as 0·5 kg/dm³ and a thermal conductivity of 0·15 W/m.degK. Such lightweight

concrete has a compressive strength after 28 days of the order of 2 MN/m²
and a drying shrinkage below 0·03%.

However, it is possible to produce compacted mixes containing a 1:2·8
ratio of cement/expanded slate which has a density of 1·35 kg/dm³ and a
compressive strength after 28 days of 28 MN/m².

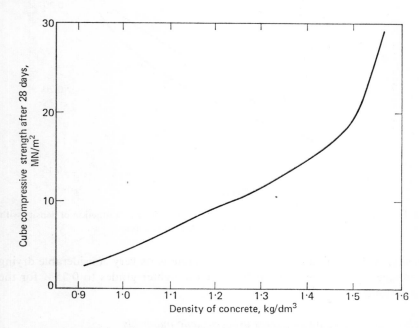

Fig. 2.3   Compressive strength of expanded pulverized fuel ash concrete
as a function of density

Pumice is a natural volcanic rock with a very spongy structure, which is
either white or off-white in colour. When mixed with cement, the concretes
produced are very similar in their properties to those produced from expanded
slate. Fully compacted mixes are, however, far weaker than those with
expanded slate as a ballast.

Scoria is a mineral somewhat similar to pumice, but it is darker in colour
and its pores are larger and more irregular. It is also used as an aggregate in
lightweight concrete.

Diatomite (kieselguhr) is derived from the skeletons of myriads of micro-
scopic plants and has a density of only 0·45 kg/dm³. It is basically hydrated
silicon dioxide. Diatomite can be expanded very considerably by heating to
1,100°C and is then mixed with cement to form various kinds of aggregate
lightweight concretes. The compressive strength of various types of diatomite
concrete is given in Fig. 2.4, which also relates the thermal conductivity with

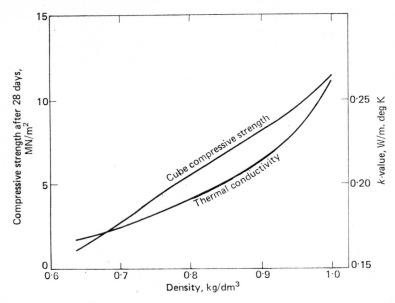

Fig. 2.4   Cube compressive strength and thermal conductivity as a function of density with respect to expanded diatomite concrete

density. A disadvantage of diatomite concrete is its very considerable drying shrinkage which varies from 0·26% for the lighter grades to 0·35% for the heavier ones.

Lightweight aggregate concretes using organic materials
Lightweight aggregate concretes are also made using a number of different organic aggregates. The most common of these materials is sawdust, which is commonly pretreated with lime and calcium chloride. Sawdust cement is used for floor finishes and precast floor tiles. One-to-one mixes of cement to sawdust can have compressive strengths of up to 33 MN/m² at a density of only 1·6 kg/dm³, but this strength falls considerably as the ratio of cement/sawdust is reduced (Fig. 2.5). Drying shrinkages are between 0·25 and 0·5% and moisture expansion can go up to 0·3% with high sawdust content types of mixes.

Lightweight aggregate concretes using particles of expanded polystyrene are also used for certain special purposes such as the construction of thermally insulating underfloor screeds.

Exfoliated vermiculite concrete
Very lightweight aggregates such as exfoliated vermiculite, expanded perlite and ultra-lightweight expanded slag are generally used only for insulating concrete units which do not perform any loadbearing function. Vermiculite is

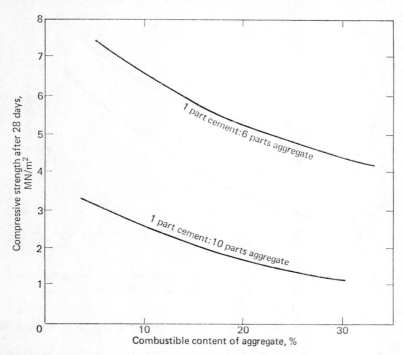

Fig. 2.5   Effect of carbon content of aggregate on strength of lightweight aggregate concrete

a hydrated magnesium silicate material which is very similar in appearance and chemical structure to mica. The material is dried to contain about 3% moisture and is then heated rapidly to a temperature between 1,000 and 1,200°C. This causes it to expand very considerably so that it swells up to 20 times its original thickness producing a material which has a density as low as 0·09 kg/dm³. Expanded perlite is obtained in the form of a light cellular material, by flash-heating particles up to 1,800°C to a density as low as 0.065 kg/dm³. By modifying the foaming process somewhat, it is possible to produce even blast furnace slag in the form of ultra-lightweight foam, with densities approaching 0·1 kg/dm³.

Concretes made from these materials have very low conductivities, but their strengths are poor also. The relationship between density, compressive strength after 28 days and conductivity can be seen from Fig. 2.6.

AERATED OR GAS CONCRETE

These materials are lightweight cellular substances which differ from other concretes in that they do not contain any aggregates. Aerated concretes are made in two ways:

D

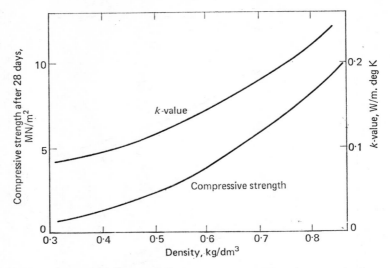

Fig. 2.6 Relationship between conductivity, density and compressive strength of concretes using exfoliated vermiculite or expanded perlite

1 By liberating a gas chemically within the concrete mix to produce a fine pore structure of non-interconnecting sacks.

2 By adding a foaming agent to the concrete mix and agitating to produce a solid foam.

Both grades can be produced in densities varying from 0·4 to 1·5 kg/dm³ and are made in two modifications, the *in situ* cast type and the precast products which are usually high-pressure-steam cured.

Production of gas within the structure of the concrete
A slurry is made from Portland cement, lime and ground silica material which is then mixed with the gasifier. Two main processes are used to produce the gases which cause the expansion of the concrete.

The most common is the use of finely powdered aluminium which reacts with lime as follows:

$$2Al + 3Ca(OH)_2 + 6H_2O \rightarrow 3CaO . Al_2O_3 . 6H_2O + 3H_2$$
$$\text{(tricalcium aluminium hydrate)}$$

An alternative method uses zinc which reacts as follows:

$$Zn + Ca(OH)_2 \rightarrow CaZnO_2 + H_2$$
$$\text{(calcium zincate)}$$

When gas concrete blocks are being made, the concrete which has been allowed to expand is then wheeled into an autoclave for steam curing, which is usually carried out at a gauge pressure of 10 bar (1 bar$=10^5$ N/m²) for

14—18 hr which corresponds to a temperature of 185°C. When Portland cement is used as a binder, the initial development of strength depends mainly on the normal setting of the cement. Autoclaving is simply used to reduce the final hardening time for the gas concrete. However, lime is often used as a binder, without any Portland cement at all. The final strength of this type of gas concrete depends entirely upon the autoclaving conditions, because the material depends primarily upon the following reaction, which only takes place with any appreciable speed under high-temperature-steam conditions:

$$Ca(OH)_2 + SiO_2 + aqua \rightarrow CaO.SiO_2 \text{ aqua}$$

Gas concrete precast units need to incorporate the siliceous matter in a very finely ground state. Pulverized fuel ash is particularly useful in this context as it contains a large percentage of SiO_2, yet needs no grinding.

PRODUCTION OF GAS CONCRETE USING A FOAMING AGENT

In this technique cement and ground silica are mixed together with water in a standard concrete mixer to form a slurry. A foaming agent such as sodium lauryl sulphate or a similar material is then added, followed by high speed mixing in a special foam generator. This produces a wide range of solid foams, which are then allowed to set and are steam cured in autoclaves in the usual way.

Properties of gas concrete
Once the gas concrete has been discharged from the autoclaves, it has become a building material (Fig. 2.7) which possesses many of the properties of

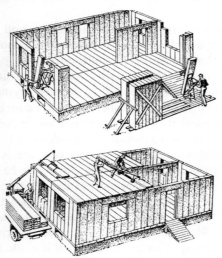

Fig. 2.7   Gas concrete used in small house construction
(By courtesy of AB Siporex, Sweden)

timber, yet is naturally vermin-proof, rot-proof and incombustible, as it is entirely mineral in nature. It has a low drying shrinkage and moisture expansion rate, the difference in dimensions between the saturated and the completely dry states amounting to less than 0·05%. Chemically speaking it has been found that the main cementing agent in autoclaved gas concrete is well crystallized tobermorite ($3CaO.2SiO_2.3H_2O$), which is present in a regular crystalline pattern. Air-cured gas concrete, on the other hand, contains mainly

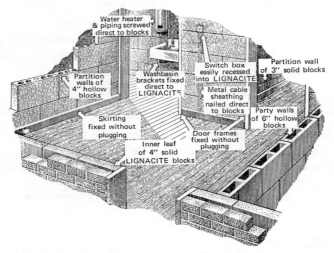

Fig. 2.8   Various uses of lightweight concrete (Lignacite) in a building
(By courtesy of Lignacite Limited)

badly crystallized tobermorite in a tobermorite gel, and is in consequence liable to much higher shrinkage when exposed to dry conditions, or growth when exposed to damp atmospheres. While autoclaved gas concrete shrinks 0·04% when dried out to a 40% relative humidity, a similar grade of air hardened gas concrete shrinks some 0·25%, because crystallized tobermorite has a low shrinkage, while gel form tobermorite has a high degree of shrinkage. The modulus of elasticity of gas concrete is low in comparison to heavy concrete. Normal concrete has a modulus of elasticity of $2—3·3 \times 10^4$ MN/m². In contrast the modulus of elasticity of gas concrete is between 1·5 and $3·5 \times 10^3$ MN/m², depending upon density. Finally, it has been found that with gas concretes made by the liberation of gases within the structure, the enclosed pore structure is quite impervious to water penetration. The writer has himself seen gas concrete blocks which have floated right across the North Sea from Norway to Scotland, and which are used by Scottish fishermen as floats to keep their nets up. Gas concrete, which is made by the addition of foaming agents, has interconnecting pores and, because of this, transmits water by capillarity. While the former grade of gas concrete can be used without detriment as a single thickness outer course in building construction, the latter type would be wholly unsuitable to be used in this way, because of its liability to rain water penetration.

Table 2.2 gives the relationship between the density, compressive strength, modulus of elasticity and conductivity for gas concrete.

TABLE 2.2

Density kg/dm³	Modulus of elasticity, MN/m²	Compressive strength, MN/m²	Thermal conductivity, W/m.degK
0·32	—	0·7	0·11
0·50	1·6	2·0	0·14
0·63	2·3	3·4	0·17
0·8	3·5	4·6	0·23

The effect of gas concrete upon embedded reinforcement
When normal dense concrete is reinforced by the inclusion of steel bars or steel mesh, corrosion is inhibited both by the close proximity of alkaline substances to the steel and also because of the exclusion of air and moisture from the steel surface. In the case of gas concrete, corrosion is bound to be more marked for the following reasons:

1 Due to the open pore structure of the material air and water can easily reach the surface of the metal.

2 The gas concrete is readily carbonated as atmospheric carbon dioxide drifts through the pore structure and reacts with free lime.

3 Steam curing itself tends to increase the rate of corrosion.

While the rate of corrosion of steel reinforcement in unpolluted air is not excessive, steel reinforcement readily corrodes when a reinforced gas concrete is exposed to an industrial atmosphere. Due to the effect of differential aeration (see Chapter 8) the rate of corrosion of embedded steel rods is actually greater than when they are exposed to the atmosphere in the open.

To prevent reinforcement rods from corroding they have to be protected. Unfortunately, none of the conventional rust prevention processes can be applied and therefore the best method has been to surround reinforcement bars by a layer of dense concrete before embedding them in the gas concrete mix. Another method which has been used with some success is to add some rubber latex and casein to the gas concrete mix.

Chemical resistance of gas concrete
Gas concrete is readily attacked in industrial atmospheres by the various acid gases such as CO_2, SO_2, SO_3, etc., which are contained in them. These, together with water, leach out free lime and other soluble salts, a process which is then followed by efflorescence, and finally cracking. Gas concrete also ages markedly if exposed to temperatures in excess of 200°C for appreciable periods of time.

It has been found that gas concretes of higher density tend to resist acid gases better.

Protection against polluted atmospheres is given to gas concrete by painting the surface with bituminous paint, which can subsequently be covered with aluminium paint. Various types of protective acrylic paints are also commonly applied to both external and internal surfaces of gas concrete, usually in the form of sand/acrylic mixes, which have an extremely good protective action against acid gases.

2.5 Asbestos and asbestos products

The commercial value of asbestos depends largely on two physical characteristics—its incombustibility and its unique fibrous structure. The latter permits it to be separated into filaments which, in the variety of asbestos most frequently used, normally possess high tensile strength and marked flexibility.

Other valuable properties are resistance to heat, moisture and corrosion. However, these properties, like some of those already mentioned, are found to vary considerably between different specimens of the same mineral.

VARIETIES OF ASBESTOS

Some thirty or more minerals of fibrous crystalline structure comprise the asbestiform group but only six have any economic significance. These are, in order of importance, chrysotile, crocidolite, amosite, anthophylite, tremolite and actinolite. Chrysotile is a fibrous form of serpentine; the other five comprise the amphibole group.

It is important to emphasize that the name 'Asbestos' given to these important minerals does not distinguish between its natural types: the two main groups being serpentine and amphibole asbestos. The mineralogical differences are a result of the varieties of matrix in which asbestos occurs. The chemical basis of all types of asbestos is magnesium silicates combined with lime or alkalis in varying amounts. An approximate chemical analysis of the two main groups is shown in Table 2.4.

SERPENTINE ASBESTOS OR CHRYSOTILE

This type is the most important industrially, and has an ideal chemical constitution of $Mg_3Si_2O_5(OH)_4$; however, its actual composition is somewhat variable and the analysis figures are shown in Table 2.4.

The colour of chrysotile varies from a pure white to greyish green, depending on the impurities present. Chrysotile remains unaffected by heat up to 450—500°C, when it begins to lose water from its structure; this process is essentially complete at temperatures of the order of 700°C, but the residue does not fuse until temperatures around 1,450—1,500°C are reached. Chryso-

TABLE 2.3

Sources, composition and characteristics	Type of asbestos		
	Chrysotile (white)	Crocidolite (blue)	Amosite
Approximate temperature at which degradation commences, °C	932	750	1,000
Approximate fusion temperature, °C	2,700	2,150	2,550
Main virtues	Excellent resistance to heat and all liquids except strong acids	Resistance to heat and very strong acids	Heat resistance
Principal products	Asbestos textiles	Asbestos textiles	Preformed slab-type thermal insulation

tile fibres are resistant to attack by alkalis, but the action of mineral acids is fairly rapid, and with strong solutions the fibres will ultimately dissolve completely.

A proportion of talc is sometimes found with this asbestos and occurs as a mixture. Some authorities believe that it is the presence of talc that gives chrysotile its peculiar slippery feeling, which incidentally makes it much easier to work than amphibole. Chrysotile asbestos occurs in fibres up to a maximum length of approximately 7·5 cm, although approximately 70% of the mined material is short and it is highly probable that the percentage of fibres 3·8 cm and over is very small.

TABLE 2.4

Constituent	Typical analysis			
	Chrysotile		Crocidolite Cape Blue	Amosite
	Canadian	Rhodesian		
	%	%	%	%
Magnesium oxide (MgO)	41·41	33·9	2·64	3·96
Silicon oxide (SiO_2)	40·49	40·9	51·64	50·24
Water (H_2O)	14·06	13·9	4·01	3·0
Ferric oxide (Fe_2O_3)	2·53	2·4	34·38	7·8
Ferrous oxide (FeO)	—	—	—	32·0
Aluminium oxide (Al_2O_3)	1·27	1·5	—	—
Calcium oxide (CaO)	—	0·2	0·05	traces
Sodium oxide (Na_2O)	—	—	7·11	2·12

Almost all the world's resources of chrysotile are associated with ultrabasic rocks, or those igneous rocks composed of ferromagnesium silicates which were injected upwards into the earth's crust. The actual manner in which asbestos was formed is a matter of conjecture but the observed facts lead to

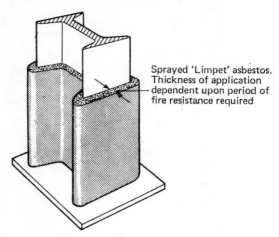

Sprayed 'Limpet' asbestos. Thickness of application dependent upon period of fire resistance required

Fig. 2.9   Steel girders covered with sprayed asbestos
(By courtesy of Turner Brothers Asbestos Limited)

the following explanation. Various changes of pressure in the earth's crust caused innumerable cracks to form in ultrabasic rocks and hot ground waters, containing dissolved mineral salts and carbon dioxide, were forced into the cracks under high pressure. As time passed, chemical reactions took place between the substances dissolved in the waters and the rocks forming the walls of the cracks, resulting ultimately in the formation of even and closely packed asbestos fibres.

Properties
The properties of chrysotile fibres which particularly adapt them to the manufacture of asbestos textiles are length, strength, toughness, flexibility and a minimum of electrically conductive particles. Chrysotile fibres are the most flexible of all asbestos fibres. Their tensile strength is high and individual values of more than 5·5 GN/m^2 have been recorded.

Recent studies under the electron microscope indicate that chrysotile fibres may be tubular in structure but with the centres of the tubes possibly filled with less highly crystalline material of the same chemical composition.

AMPHIBOLE ASBESTOS

The types of asbestos in this group differ from chrysotile asbestos in containing a larger percentage of silica, less magnesium, and larger amounts of iron, aluminium, sodium and calcium.

Slight differences from these figures do, of course, occur. It was popularly thought, a few years ago, that blue asbestos was inferior in its heat-resisting properties to chrysotile, but contemporary use of Cape Blue for heat insulation has shown that it is quite good. On making an examination of amphibole asbestos, it will be noticed that it is more harsh and not as slippery to the touch as chrysotile, and these properties make it more difficult to work. Crocidolite is the blue asbestos of commerce, it is more difficult to process into a spinnable fibre than chrysotile but its tensile strength is generally very high. The most important characteristic of crocidolite, however, is its superior resistance to attack by acids.

AMOSITE

Like crocidolite, amosite contains approximately 35% iron expressed as oxides, but mainly in the less oxidized, ferrous state. It contains greater percentages of magnesia (up to 6·5) and alumina (up to 6·0) and the alkali metals are present only in relatively small amounts. The combined water content is similar to that of crocidolite.

Amosite asbestos possesses good tensile strength and in certain applications its resistance to heat is superior to that of chrysotile or crocidolite.

Amosite fibres are used principally in the manufacture of thermal insulation materials.

Fibres of amosite as long as 30 cm are not uncommon; amosite has, however, a more limited range of uses than chrysotile and crocidolite, because of its relatively low strength and its lack of flexibility.

The composition of fibre and location of main deposits is given in Table 2.5.

TABLE 2.5

Variety	Chemical composition	Qualities	Important deposits
CHRYSOTILE	Hydrated magnesium silicate	Flexible and heat resisting (less acid-proof)	Canada Southern Rhodesia Russia (USSR)
AMPHIBOLE CROCIDOLITE	Silicate of iron and sodium	Acid-resisting, tensile strength	South Africa
Amosite	Silicate of iron and magnesium	Brittle, long fibres	Transvaal
Anthophyllite	Silicate of magnesium and iron	Brittle, but sometimes silky	USA Africa
Actinolite	Silicate of magnesium, calcium and iron	Brittle, but more acid-proof than chrysotile	North America
Tremolite	Silicate of calcium and magnesium	Brittle, but more acid-proof than chrysotile	Italy Africa Balkans America

Asbestos itself cannot be classed as a refractory. Normally its qualities are sufficient for it to stand the temperature of superheated steam or the melting-point of the commonly used metals and alloys. In addition to its ability to withstand fairly high temperatures it has the valuable property of being a poor conductor of heat. The low heat conductivity is probably due to its cellular construction; at the same time it may have some bearing on its electrical insulation properties.

A rough outline of the uses of different fibre lengths is as follows:

Powder, and below 6·5 mm	Asbestos cement sheets, asbestos paper and millboards, moulded plastics, components, fireproof paints, etc.
6·5 to 19 mm	Spray work
13mm upwards	Textile fibres for carding, spinning and weaving

MANUFACTURE

The primary processes occur in two stages, crushing, and 'opening' or fiberizing.

Crushing

It should be noted that with some of the softer fibres crushing is unnecessary although in practice the bulk of material is crushed. This often takes place in an edge mill grinder which consists of a heavy granite-based pan of circular shape and two frictionally driven granite rollers. A cruder version of this type of mill is frequently seen in use grinding mortar for building work. A charge of asbestos crude is placed in the mill and ground for a given time. Too much crushing destroys the fibre, making it limp and inert. If too little time is allowed, then correct fiberization is not achieved and the product is 'spiky'. The material is then removed from the crusher, and blended, if necessary, with other grades of asbestos.

Opening

The material is fiberized in a high speed mill that has a steel case with openings in the periphery, and rotating inside the case are beaters constructed of steel. In this mill the asbestos is flung against the openings and then forced through them. The fiberized material is then carried away into a collecting apparatus and bagged. For fiberizing the softer grades of asbestos, a mill with open sides is sometimes used, the fibre passing completely round the machine and being blown through a tangential opening into the collecting apparatus. A further type of machine used for opening the fibre consists of a rotating cage of robust construction which carries a number of pivoted blades. When the cage rotates, these blades or beaters assume a radial position due to centrifugal force and thus beat the fibre in a similar manner to the previously mentioned type of mill. This type of machine usually has perforated plate screens and is generally used for the shorter classes of fibre. The pivoted beaters make for a softer operation in that the machine cannot jam if a piece of rock or metal is inadvertently fed into the inlet.

Fibres intended to be used on textile work receive a more exacting treatment in order to eliminate uneven lengths and off pieces of rock or unopened strands of fibre. After crushing, the material is screened in a machine of the vibrating type. This operation removes the main body of waste, such as rock and short fibres. The material is then fiberized ready for the textile operators. Extreme care is taken to eliminate particles of rock.

The short asbestos fibres which constitute the bulk of the mined material are utilized in the manufacture of boards and moulded articles (Fig. 2.10).

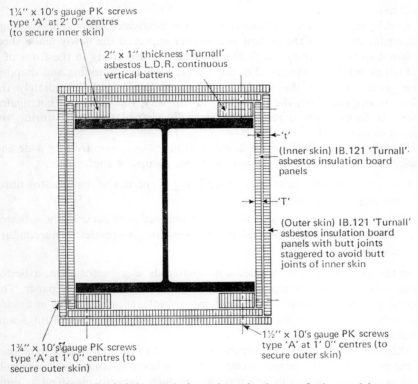

1¼" x 10's gauge PK screws
type 'A' at 2' 0" centres
(to secure inner skin)

2" x 1" thickness 'Turnall'
asbestos L.D.R. continuous
vertical battens

←'t'

(Inner skin) IB.121 'Turnall'·
asbestos insulation board
panels

←'T'

(Outer skin) IB.121 'Turnall'
asbestos insulation board
panels with butt joints
staggered to avoid butt
joints of inner skin

1½" x 10's gauge PK screws
type 'A' at 1' 0" centres (to
secure outer skin)

1¾" x 10's gauge PK screws
type 'A' at 1' 0" centres (to
secure outer skin)

Fig. 2.10   Steel girders made fire-resistant by the use of asbestos slabs
(By courtesy of Turner Brothers Asbestos Limited)

This material is used for a great variety of structural purposes, such as the fabrication of roof and wall cladding, hutting, partitions and ceilings. In addition moulded asbestos/cement products are available as rain water and soil goods, flue pipes, cisterns, cable conduits and troughs. Other important uses are in structural fire protection, thermal insulation and sound absorption treatments.

In the mixing of fibre with cement, a very considerable excess of water is introduced. This may appear to jeopardize the setting properties of the cement, but is a purely temporary phase employed primarily for the proper positioning of the fibre. Excess water is rapidly drained off, leaving a very thin film or lamina of cement with embedded fibre on an endless belt which is eventually carried to a revolving forming bowl or mandrel. This forming bowl picks up film after film in a continuous operation, superimposing one on the other to a predetermined thickness; at this point, the fabric is removed for further processing. Fundamentally, this is the basis of all asbestos/cement products whatever their ultimate shape or design and, except in the production of pipes, the material at first appears in the form of a flat sheet of uniform thickness.

At this stage, the fibres' reinforcing effect is clearly demonstrated. Although the setting agency of the cement has not yet appeared the newly made sheet is found to be extremely tough and may be handled freely in the form of a wet felt or blanket with several hours still available for moulding and shaping. The greater part of this work is, of course, performed immediately the fabric is stripped from the bowl and, in this way, the variety of corrugated sheets is formed which remain on specially shaped templates during the period of initial set.

The present-day employment of asbestos in industry is extremely wide and varied but can be divided into two principal groups of application:

1 One in which the heat- and fire-resisting properties of the asbestos fibres are of prime importance.

2 One in which the asbestos fibres are combined with cement, the asbestos acting as a form of reinforcement. Fire-resisting properties are a secondary consideration.

In the first group are included such materials as asbestos yarn, asbestos cloth, asbestos rope for pipe wrappings, mill-board and asbestos paper. The second group incorporates asbestos/cement sheets, tiles and slates in a wide variety of form and colour, and an extensive range of pipes, fittings and moulded products.

Although the fire-resisting properties are not of prime importance in respect of the products in the second group, in cases where the constituent materials used in the manufacture of any product are each correctly classified as non-combustible, then the final product is essentially non-combustible.

The British Standard Specification 476/1953, 'Fire Tests on Building Materials and Structures', test is applied to materials normally used in the construction of a building, in order to classify their behaviour when subjected to comparatively high temperatures.

Six specimen pieces of the material to be tested are heat-dried at a temperature of 100°C for a period of 6 hr and then allowed to cool off in a dry atmosphere. Each specimen is then inserted in turn and left for 15 min in a specially constructed furnace, the temperature of which has been stabilized at 750°C.

The material is considered to be combustible, if during the test period, any one of the six specimens:

1 Flames.
2 Produces vapours which are ignited by the pilot flame.
3 Causes the temperature of the furnace to be raised 50°C or more above 750°C.

Asbestos/cement products normally offered for general building and other constructional work can broadly be classified as:

1 Low density or insulation boards.
 a Ships board.
 b Insulation board.
2 Ordinary density or semi-compressed material.
3 High density or fully compressed material.
4 Corrugated products.

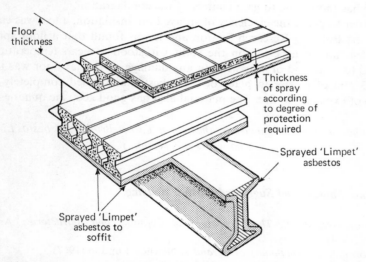

Fig. 2.11 Asbestos sprayed onto the soffit and load-bearing beams of a building
(By courtesy of Turner Brothers Asbestos Limited)

SPRAYED ASBESTOS

If a microscopical examination of asbestos fibre is undertaken, it will be revealed that each strand is composed of many finer pieces adhering together, in fact, the diameter of well separated asbestos fibre is in the region of 0·5 microns (5×10^{-3} mm). Further detailed examination will reveal that an inch cube of asbestos contains 15 million miles of fibre. There are tiny longitudinal spaces between the fibres in each bundle. These spaces act as capillary tubes and draw in liquids. Thus, if the fibre is intimately mixed with a certain quantity of liquid, a suction effect is created in each bundle of fibres. During the last two decades, a practical application of this effect has been found for the spraying of asbestos and other fibrous materials (Fig. 2.11).

HEAT INSULATION

Asbestos is recognized as a protection against fire, and spraying, being an ideal method of applying the material, is extensively used in industry. In many buildings where a fire hazard exists, it is customary to spray the structural steel work with a given thickness of fibre, the thickness, of course, being dependent upon conditions. For further protection against fire, the underside of floors can also be treated, thus preventing the penetration of flames to other rooms.

In case of fire, the steel framework of the building, being efficiently insulated, remains reasonably rigid, i.e. the yield point is not reached and the Fire Service has more time to gain control of the conflagration.

In order to prove the efficiency of sprayed fire insulation, a test was carried out at the Building Research Station and it was found that a floor sprayed with 25·4 mm of asbestos on the soffit withstood a mean temperature of 1,000°C for a period of 4 hr without any damage. A similar floor was tested with 13 mm of plaster in place of the asbestos and failed completely after 25 min of exposure to fire. The material also has good acoustic qualities.

(The author herewith acknowledges the help of Turner Brothers Asbestos Limited who wrote this section.)

Literature Sources and Suggested Further Reading

1. BUDNIKOV, P. P., *The Technology of Ceramics and Refractories*, Arnold, London (1964)
2. DODD, A. E., *Dictionary of Ceramics*, Newnes, London (1967)
3. GYPROC LIMITED, *The Gyproc Book* (1968)
4. *Minerals Yearbook*, US Department of the Interior, Bureau of Mines (1967)
5. RUDNAI, G., *Lightweight Concretes*, Akademai Kiado, Budapest (1963)
6. SHORT, A., and KINNIBURGH, W., *Lightweight Concrete*, CR Books, London (1963)
7. VARLEY, E. R., *Vermiculite*, Wiley, New York (1966)
8. TURNER BROTHERS ASBESTOS LIMITED, Technical Books and Brochures

Chapter Three   Ceramics

The term 'ceramics' describes all products made from burnt clay. Most ceramics are mixtures of clay with other materials such as feldspar ($Na(K)AlSi_3O_8$), silica (SiO_2), sillimanite ($Al_2O_3 . SiO_2$) and many other substances.

Ceramics are widely used in many industries. In the building and construction industries they are mainly used in the manufacture of the following:

1 Bricks, including numerous varieties from the cheapest common bricks used for internal wall leaves, to high grade engineering bricks used for the construction of machinery bases.

2 Roof tiles.

3 Floor tiles, both unglazed and glazed.

4 Wall tiles of all types, including unglazed brickwork tiles, and whiteware tiles, which are usually glazed. Decorative coloured and glazed external tiles are also included in this category.

5 Sanitary fittings of all kinds, usually known as whiteware, which are glazed either white or with a coloured finish.

6 Refractory ceramics for furnace and flue linings.

3.1 Clays

The raw materials of ceramic manufacture are a variety of different clays. Clays are natural products which were formed many years ago by the weathering of granite and gneiss which constitute the main part of the earth's surface. Feldspar, which is a mixture of $K_2O . Al_2O_3 . 6SiO_2$ and $Na_2O . Al_2O_3 . 6SiO_2$, is the most common of these materials and is weathered according to the following reaction:

$$K_2O . Al_2O_3 . 6SiO_2 + 2H_2O + CO_2 \rightarrow Al_2O_3 . 2SiO_2 . 2H_2O + 4SiO_2 + K_2CO_3$$

The potassium carbonate, which is water soluble, is leached out, leaving the remaining materials behind. They are usually mixed with a large number of other materials, particularly ferric oxide (Fe_2O_3) which is responsible for the red coloration of many clays, and limonites (hydrated iron oxides) which impart a buff coloration. Many clays, in addition, contain a good deal of organic matter and also titanium oxide. Apart from kaolinite ($Al_2O_3.SiO_2.2H_2O$), which is the most common ingredient, clays also contain some montmorillonite $(Mg, Ca)O.Al_2O_3.5SiO_2.nH_2O$ and illite which can best be written as:

$$n_1K_2O.n_2MgO.n_3Al_2O_3.n_4SiO_2.n_5H_2O$$

Clays are subdivided into the following main groups.

Kaolins
These are refractory clays consisting mainly of pure kaolinite, with a negligible quantity of impurities. They form translucent grades of chinaware. Typical samples of such clays contain about 38% Al_2O_3, 45% SiO_2, 14% bound water with about 1% Fe_2O_3 and 1·5% TiO_2, the remaining 0·5% being accounted for by other materials.

Ball clays
These contain a fair proportion of organic matter, and are, in consequence, dark in colour when mined. They have good plasticity, an excellent dry strength, as well as a long vitrification range. These materials, which are fairly light after firings, are used for lower grade whitewares. A typical ball clay has 32·5% Al_2O_3, 52% SiO_2 and 1·5% Fe_2O_3 with an ignition loss of about 9% which accounts for both combined water and organic materials.

Terracotta clays
These contain appreciable percentages of iron compounds and are reddish or buff coloured after firing. They are the main types of clays used for the manufacture of bricks, roof tiles and other building components. These clays are by far the most widely distributed.

Fire clays
Fire clays are generally not white-burning as they contain 2—3% Fe_2O_3. They have a low colloidal content and in consequence a poor plasticity, in the fluid state. Their advantage lies in the fact that, when they have been burnt, they are able to withstand high temperatures without slagging.

Stoneware clays
These clays contain broken down feldspar, which reduces the fusing temperature and results in the formation of a very dense and partially vitrified product.

3.2 Structural clay products

The term 'structural clay products' includes common bricks, facing bricks, industrial bricks, decorative bricks which may be glazed, as well as such auxiliary components as chimney pots, flue linings, etc. Numerous other items are also included under the same heading, such as tiles of all description, sewer bricks, clay conduits and hollow blocks.

All these products are made from clays, or mixtures of clay and shale. Many of these contain up to 5% Fe_2O_3 as well as quantities of manganese dioxide. In the formulation, clays that contain appreciable amounts of $CaCO_3$ or $CaSO_4$ are avoided. It is common to mix raw clay with ground up broken pottery and other fired products. In some cases the percentage 'grog', as this is called, amounts to as much as 25%.

Three brick-making processes are commonly used, called, respectively, the soft mud process, the stiff mud process and the dry process. In the latter the clay, which contains less than 10% moisture, is pressed into moulds at a pressure of about 100 bar.

The first stage is the drying of the products, followed by firing. Mechanical dehydration is complete when the temperature of the bricks has been raised to 150°C and between 400—600°C chemical dehydration begins. Oxidation of organic and other materials takes place between 250°C and 700°C. The ultimate firing temperature depends upon the intended use of the final brick. In general, the higher the final firing temperature of the brick, the lower the porosity of the product and the greater its strength.

The normal firing temperatures employed are as follows:

Common brick	970—1,150°C
Facing bricks and roof tiles	1,000—1,240°C
Sewage pipes	1,030°C
Engineering bricks	1,200—1,300°C

Clays are either red-firing or buff-firing. When red-firing clays are heated, light shades of pink appear at first. In an oxidizing atmosphere a deep red colour is developed. When alternating oxidizing and reducing conditions of firing, buff-firing clays give yellow and brown colours, while red-firing clays turn dark brown.

Mechanical properties of bricks

The strength of a brick usually depends upon its porosity and hence on the degree of vitrification attained. Its mechanical properties are related only to a relatively minor extent to the nature of clay used.

The following general formula applies to the compressive strength of standard clay bricks:

$$S_c = A(x/y)^3$$

where S_c (MN/m²) is the compressive strength of the brick, A a constant depending on the nature of the clay used and the degree of firing employed, and which is usually of the order of 70—100 MN/m², x (kg/dm³) the density of the brick and y (kg/dm³) the density of the brickwork.

E

The porosity of a brick is expressed by the related formula

$$p \text{ (porosity)} = [(y-x)/y] \times 100\%$$

The common porosities found are as follows:

Type of brick	Density, kg/dm^3	Porosity, %
Low-grade common brick	1·2	55
Medium-grade common brick	1·6	41
Low-grade facade wall brick	1·8	33
Better-quality facade wall brick	2·0	26
High-grade external brick	2·3	15
Grade B engineering brick	2·5	7·5
Grade A engineering brick	2·6	3·7

The compressive strength of a good common brick is around 35 MN/m², of a facade brick around 40—42 MN/m², and 50 and 70 MN/m² for grade B and grade A engineering bricks, respectively.

Compressive strengths, tensile strengths and shear strengths of bricks vary very considerably with the grade of brick used, whether the brick is taken as a whole or broken prior to testing. The moisture condition of the brick is also of importance and there can be variations of up to 50% within a single batch.

For most samples of bricks the tensile strength is seldom more than 10% of the compressive strength, and often amounts to as little as 5%. The shear strength of bricks varies from 25 to 30% of the compressive strength, while the flexural strength (strength resisting twisting) can amount to as much as 40% of the compressive strength.

In general, some 80% of the number of pores in bricks are inter-connecting so that during prolonged immersion in water a very considerable absorption of water takes place. Bricks are very resistant indeed to atmospheric impurities and are unaffected by acids, although they are affected by alkalis. Soluble salts in the clay, which remain in the final brick, are the cause for brick 'efflorescence'. During periods of heavy rainfall such salts, usually calcium sulphate, magnesium sulphate, sodium sulphate and various nitrates, are washed out of the brick, and appear subsequently as a white sparingly soluble scum on the brick surface.

The trouble mainly arises with bricks which have not been fired too well. It can be prevented by mixing a slight excess of barium carbonate with the clay which then reacts with any sulphate ions present to form insoluble barium sulphate:

$$BaCO_3 + MgSO_4 \rightarrow BaSO_4 + MgO + CO_2$$

Highly porous bricks, with a large percentage of open pores, have poor resistance to freezing. For this reason porous bricks must never be used for facade purposes since water, which penetrates major pores, expands upon freezing and causes the bricks to break up (spall).

Water resistance of brickwork

As nearly all bricks that have been burnt to lower temperatures are porous with some 80% of the number of open pores, bricks resist driven rain far less than mortar, concrete and other building materials. This includes, surprisingly enough, even most grades of lightweight concrete mentioned in Chapter 2. Due to the fact that the angle of capillary contact of brick is close

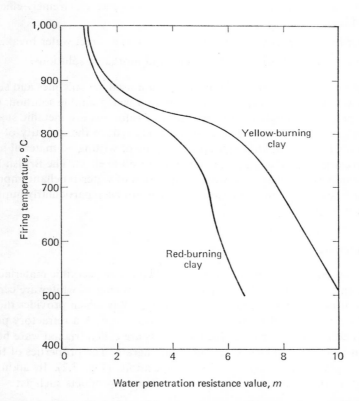

Fig. 3.1   Relationship between firing temperature and water resistance of brick

to zero (the capillary force $= F\cos\Theta$ where Θ is the angle of capillary contact and F the surface tension force), there are strong capillary forces which push water through its pores. The rate of passage of water through a brick varies enormously with the grade. Swedish workers have developed the definition of a water resistance figure m where a brick is considered to have an m-value of 1 if a 25 cm thick wall, which is exposed to water on one side, is completely penetrated by moisture in 10 hr. The m-value varies considerably with the firing temperature, and some values are given in Fig. 3.1.

Methods of water-proofing brickwork

The main methods used are the following:

1 Rendering of brickwork either on outside or between courses with a concrete mix, which has a water penetration figure which is a small fraction of that of brickwork (between 2 and 10%).

2 Insertion of plastic or other film between bricks or on top of them to stop water penetration. Many plastic foams operate extremely efficiently in this respect.

3 Insertion of a cavity between bricks to act as a direct water break.

4 Impregnation of bricks with organic and inorganic solutions.

The most common methods of impregnation used are silicones and siliconates, as well as sodium silicate in the form of a water soluble solution. Other water-repellant materials used in the form of solutions are metallic stearates dissolved in hydrocarbons. These considerably reduce the quantity of water penetrating brickwork, although up to the time of writing no material has yet been developed which has a permanent and absolute effect. The British Building Research Station has dealt with the question of water-repellant impregnations for brick fairly comprehensively and has not been particularly impressed by their general utility.

3.3 Whitewares

This term is commonly applied to glazed and unglazed ceramic materials with a white and fine texture. The clay used for the mixing of whiteware ceramics consists of plastic or colloidal portion of clay, a flux which provides the glass bond during firing and is mostly feldspar, together with a refractory portion which is either quartz or flint. Such materials are called triaxial ware because they can be represented on a Roozeboom diagram. The properties of triaxial ware are governed rigidly by their compositions (Fig. 3.2). In addition, a number of other materials are used for whiteware products such as:

Corderite	$2MgO . 2Al_2O_3 . 5SiO_2$
Forsterite	$2MgO . SiO_2$
Steatite	$MgO . SiO_2$
Zircon	$ZrO . SiO_2$
Mullite	$3Al_2O_3 . 2SiO_2$

The higher the temperature of firing the lower the degree of porosity, because of greater vitrification. Most whiteware products are fired at temperatures above 1,310°C which vitrifies the main body. In the case of china production, the object is then cooled down to room temperature and refired at 1,180°C. For porcelain manufacture a firing temperature as high as 1,390°C is used.

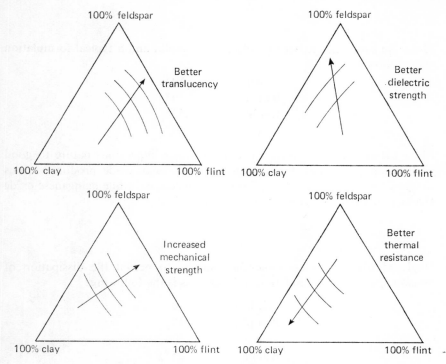

Fig. 3.2   The properties of triaxial whiteware

Sanitary ware is mostly single fired and is prepared from material with the following approximate composition:

Feldspar	30%
Ball clay	20%
Kaolin	25%
Quartz	25%

As the units are large, both the drying cycle and the firing cycle are carefully controlled. Sanitary ware is glazed, using a mix which contains the following molecular proportions:

$$K_2O = 0.20; \quad Na_2O = 0.05; \quad CaO = 0.53; \quad MgO = 0.15;$$
$$BaO = 0.07; \quad Al_2O_3 = 0.60; \quad SiO_2 = 5.0$$

The glaze is applied in the form of a prereacted suspended colloid, which is painted upon the surface followed by the heating of the entire unit to about 1,300°C, when the glaze materials form a thin layer of glass, fused to the main ceramic body. Coloured sanitary ware is made by the incorporation of various coloured oxides into the glazes, e.g. oxides of chromium, nickel, cobalt, manganese, etc. (See also p. 68.)

FLOOR TILES

These are somewhat richer in feldspar and kaolin and a typical formulation is the following:

Feldspar	50%
Ball clay	10%
Kaolin	30%
Quartz	10%

Floor tiles are usually unglazed, but owe their impervious nature to good vitrification techniques. Small amounts of chromic oxide produce various shades of green, cobalt oxide produces blue colours, while manganese oxide gives colours between cream and dark brown.

WALL TILES

Unglazed tiles are often somewhat porous, to permit the dissipation of moisture. A typical formulation for such tiles is the following:

Feldspar	10%
Pyrophyllite	10%
Talc	30%
Kaolin	25%
Ball clay	10%
Quartz	15%

Wall tiles are usually fired at a much lower temperature than floor tiles. Glazed tiles are produced in a way similar to that employed with sanitary ware, except that the glaze is often applied before the initial firing stage, so that the production can be carried out in one process.

CERAMICS FOR ELECTRICAL PURPOSES

These ceramics are classified into low-voltage and high-voltage types and their properties are summarized in Table 3.1.

TABLE 3.1

	Low voltage	High voltage
Water absorption, %	0—1·5	0—0·10
Density, kg/dm³	2·4	2·4
Flexural strength, MN/m²	42	73
Tensile strength, MN/m²	28	40
Compressive strength, MN/m²	240	340
Moh's hardness	7·0	7·0
Softening-temperature, °C	1,400	1,300
Dielectric strength, mV/mm	10	10

Electrical porcelains are usually rich in quartz and are fired to temperatures well above 1,300°C. Glazing is carried out by single stage firing with a high compression glaze.

REFRACTORIES

These are ceramic products that are able to withstand high temperatures, and in addition they must also be able to resist some other conditions, such as attack by slag, abrasion, mechanical and thermal shocks, etc. There are some industrial refractories on the market, such as hafnium carbide, tantalum carbide and zirconium carbide, which are able to withstand temperatures of over 3,500°C.

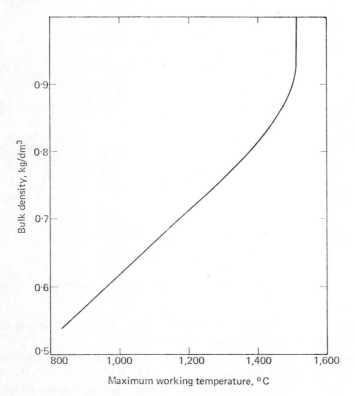

Fig. 3.3   Relationship between bulk density of fireclay brick and maximum working temperature

The types of refractories used by the building and construction industry for such purposes as furnace linings, flues, etc., are almost invariably made from fireclay. This has the general formula $Al_2O_3.2SiO_2.2H_2O$ corresponding to 39·5% aluminium oxide, 46·5% silica and 14% water. The fireclay is shaped as required, dried in a tunnel drier and then fired. Fireclays are usually rather porous with bulk densities well below 1 kg/dm³. Fireclays with a relatively high density are used where the main purpose of the refractory is to resist the flames, whereas the low-density fireclays, which also have rather poor strengths, are used for such purposes as flue linings or in conjunction with higher-density refractories for insulation purposes.

Properties of fireclay refractories
Fireclays have cross breaking strengths varying between 6·3 and 9·8 MN/m² depending upon the temperature of firing, the highest values being obtained

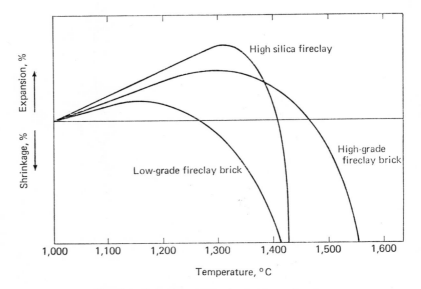

Fig. 3.4   Behaviour of fireclay bricks on heating

with fireclay which has been kept for 5 hr at 1,500°C. Young's modulus similarly varies between 12·6 and 30 GN/m².

The crushing strength of fireclay bricks increases as the clay units get heated and becomes a maximum at around 1,000°C. After this the strength falls off until it reaches a zero value at the slagging temperature, which ranges from 1,150 to 1,500°C for different types of bricks. Fireclay bricks have good spalling resistance under conditions of thermal shock, provided the percentage silica contained is not excessive. Pure silica has a considerable volume change over the temperature range 200 to 350°C and in consequence must be heated up extremely slowly over this range. The thermal conductivity of fireclay depends upon its void content and is given at different temperatures in Fig. 3.5. The thermal expansion of fireclay bricks on heating them from

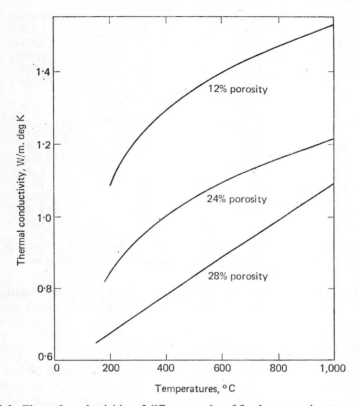

Fig. 3.5 Thermal conductivities of different grades of fireclay at varying temperatures

room temperature up to 1,000°C varies from 0·4 to 0·6% depending on grade. The rate of expansion is virtually linear with most types. After the fireclay bricks have been heated to temperatures above the slagging temperature, they contract due to the collapse of the pores (Fig. 3.4).

SEGER AND ORTON CONES

These are small cones made from carefully controlled ceramic mixes, which have the property of collapsing over a given range of temperatures. Whenever ceramics or refractories are being fired these cones are introduced at different positions in the furnace and give an indication of the temperature prevailing there. Both the Seger cones, used in Europe, and the Orton cones, used in the USA, are numbered, each number indicating a certain temperature. The following temperatures are given for Orton cones when the tip has tilted over to touch the base.

Cone number	Temperature, °C	Cone number	Temperature, °C
1	1,136	16	1,455
2	1,142	17	1,477
3	1,152	18	1,500
4	1,168	19	1,520
5	1,177	20	1,542
6	1,201	21 and 22 not used	—
7	1,215	23	1,586
8	1,236	24 and 25 not used	—
9	1,260	26	1,589
10	1,285	27	1,614
11	1,294	28	Not used
12	1,306	29	1,624
13	1,321	30	1,636
14	1,388	31	1,661
15	1,424	32	1,706
		33	1,732

For Seger cones it is not possible to give a similar table, because Seger numbers not only indicate the temperature of firing but depend on the duration.

For example, Seger cone 14 goes down at 1,410°C on rapid heating, but equally well at a temperature of only 1,390°C if the heating rate is slow.

Literature Sources and Suggested Further Readiug

1. BONNELL, D. G. R., and BUTTERWORTH, B., *Clay Building Bricks of the United Kingdom* (Ministry of Works: National Brick Advisory Council), Paper 5 (1950)

2. BURKE, J. E., *Progress in Ceramic Science*, 3 volumes, Pergamon Press, London (1962)

3. BUTTERWORTH, B., *The Properties of Clay Building Materials*, 'Ceramics, a Symposium', British Ceramic Society (1953)

4. BUTTERWORTH, B., 'Efflorescence and Staining of Brickwork', *Brick Bull.*, Vol. 3, No. 5 (1957)

5. BUTTERWORTH, B., and SKEEN, J. W., 'Experiments on the Rain Penetration of Brickwork', *Trans. Brit. Ceramic Soc.*, Vol. 61, No. 9 (1962)

6. CLEWS, F. H., *Heavy Clay Technology*, British Ceramic Society (1955)

7. DALE, A. J., *Modern Ceramic Practice*, Maclaren, London (1964)

8. 'Glazing Clay Products', *Brit. Clay Rec.*, Vol. 117, No. 37 (1950)

9. HEDGES, P. E., 'Crystalline and Glassy Phases in Commercially Fired Brick, *Bull. Am. Ceramic Soc.*, Vol. 40 (1961)

10. HOVE, J. E., and RILEY, W. C., *Ceramics for Advanced Technologies*, Wiley, New York (1965)

11. KINGERY, W. D., *Introduction to Ceramics*, Wiley, New York (1960)

12. KRIEGEL, W. W., and PALMOUR, H., *Mechanical Properties of Engineering Ceramics*, Interscience, New York (1961)

13. SEARLE, A. B., and GRIMSHAW, R. W., *The Chemistry and Physics of Clay*, Benn, London (1959)

14. SINGER, F., and SINGER, S. S., *Industrial Ceramics*, Chapman and Hall, London (1963)

15. STEWART, G. H., *Science of Ceramics*, 3 volumes, Academic Press, London (1967)

16. WAYE, B. E., *Introduction to Technical Ceramics*, MacLaren, London (1967)

Chapter Four Glass

Glass is fundamentally a mixture of complex silicates. It does not crystallize on cooling, but remains in the form of a very viscous supercooled liquid. Because of its liquid nature, glass can readily be moulded when hot, and there are numerous complex processes used for the many specialized glass products of commerce. As far as the building and construction industries are concerned, the main interests are the following:

1 *Sheet glass and plate glass* for windows including sealed cavity double glazing, Vita glass, coloured and opaque glasses, etc.

2 *Glass sections* such as shaped transparent bricks, architectural ornaments and electrical insulators.

3 *Fibre glass* used for thermal insulation purposes and for the reinforcement of plastic sections, fire-resistant curtains, etc.

4 *Surface glass films* used for vitreous enamelling and for the glazing of ceramic products.

4.1 Manufacture of glass

Glass is made by heating a mixture of silica sand and various metallic oxides together in a furnace. Owing to the high temperatures necessary, and also on account of the highly corrosive nature of the molten glass constituents, the quality of the refractory linings of the glass furnace must be very high. In practice mullite (high alumina fireclays) refractories are the ones usually used. The glass furnaces themselves are large enclosed tanks heated by gas or oil. The glass components are weighed out and the following materials are added:

1 *Cullet*, which is glass left over from a previous batch because it was not perfect.

2 *Oxidizing agents* such as manganese dioxide, which have the task of decolorizing the glass. Traces of iron are always present in glass mixes due to abrasion from mechanical handling equipment. When these traces

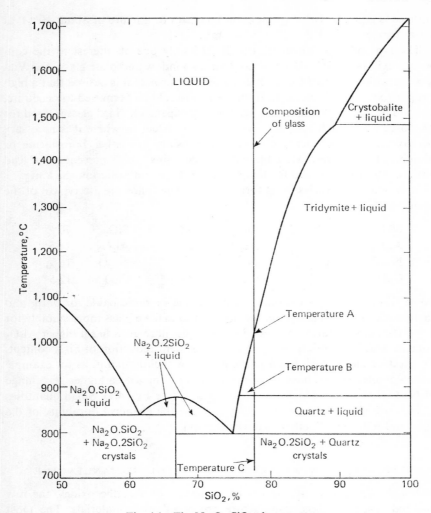

Fig. 4.1 The Na_2O-SiO_2 phase system

of iron are in the form of ferrous oxide FeO, they tend to give glass a pale green coloration. On oxidation ferric oxide is produced, which is very much paler.

3 *Various fining agents* are added, which help to eliminate the bubbles which are always initially present in the hot glass mix.

After the glass has been heated for some time to form a smooth and bubble-free melt, it is drawn off at a temperature of about 1,500°C. When it is allowed to cool slightly from this temperature, it becomes more viscous and can then be shaped. The viscosity of glass when it first emerges from the furnace is equal to 100 P (poise), but increases to a figure of about 10^{15} P at room temperature (1 P $= 10^{-1}$ N.sec/m^2).

The type of glass which is virtually the only one of interest to the construction industry is soda glass, used for all window and plate glass. The Vita glasses, which are used for specialized purposes where it is desired that a high proportion of ultraviolet light from the sun should be permitted to penetrate, contain nearly 100% SiO_2. Vita glass is very expensive. Hard glasses, used for purposes where high temperatures must be resisted, or where it is necessary to have a low coefficient of thermal expansion, are either borosilicate or aluminosilcate glasses, while optical glass contains a high percentage of lead oxides. All these are used for lamps, neon tubing and radiation shielding.

Soda/lime glass varies in its formulations. The following are typical of the many used.

SiO_2	71·5%	SiO_2	74%	SiO_2	70%
Al_2O_3	1·5%	Al_2O_3	2%	Al_2O_3	1·5%
Na_2O	14·0%	Na_2O	16%	Na_2O	16%
CaO	13·0%	CaO	8%	CaO	12·5%

Some formulations also contain barium oxide or boric oxide, or both, and there are individual changes in formulation to make a glass more suitable for a specific manufacturing process. For example, fibre glass has a higher Al_2O_3 content which can reach 14% with an equivalent reduction of SiO_2 content. To produce glass which has a particularly low melting-point, as for example when complex shapes have to be made from it, one uses a higher percentage of Na_2O than usual, which can go up to 22% (see Fig. 4.1). Small quantities of K_2O are also added, because they increase the chemical durability of the glass and reduce its liability to devitrification.

NATURE OF RAW MATERIALS USED IN GLASS MANUFACTURE

Although the glass itself consists of a mixture of metallic oxides, the raw materials used in glass manufacture are often other materials. The most common raw materials, and the oxides they form, are given in the following table:

Raw material	Glass-making oxide
Sand	SiO_2
Na_2CO_3	Na_2O
Na_2SO_4	Na_2O
$2K_2CO_3.3H_2O$	K_2O
$CaCO_3$	CaO
$Na_2B_4O_7.10H_2O$ (borax)	B_2O_3 and Na_2O
$Na_2O(K_2O).Al_2O_3.6SiO_2$ (feldspar)	SiO_2, Al_2O_3 and Na_2O (K_2O)
$CaCO_3.MgCO_3$ (dolomite)	CaO and MgO
$BaCO$	BaO

4.2 Devitrification of glass

All glasses, when they cool from the melting-temperature, pass through a range of temperature within which the liquid is likely to form one or more crystalline compounds. It is at the start of the devitrification temperature range that the free energy of the crystalline state of glass is less than the free energy of the liquid state. However, if the glass is cooled without crystalliza-

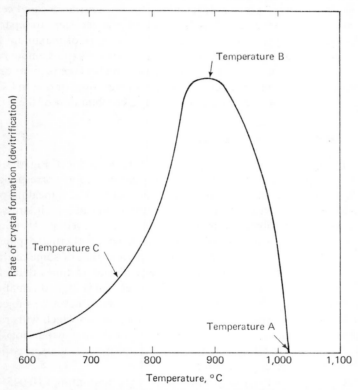

Fig. 4.2 Rate of devitrification of glass as a function of temperature

tion taking place, the increasing viscosity of the glass makes it increasingly difficult to devitrify spontaneously. Figure 4.2 shows how the rate of crystal growth is related to temperature. For a given glass, the rate of crystal growth is zero above the crystallization temperature A. Between temperature A and temperature B the rate of crystal growth increases rapidly, as the phase equilibrium favours the presence of the solid rather than the liquid state. Between temperatures B and C the crystal growth falls off drastically to become virtually zero at atmospheric temperatures. This is due to the increasing viscosity of the glass, which makes phase changes harder to take place. For each kind of glass there is a temperature at which devitrification proceeds at the maximum rate. For example, a typical soda glass with 72% SiO_2, 16% Na_2O and 12% CaO has a liquidus point at 1,010°C. The temperature of maximum crystal growth is 955°C at which temperature the rate of crystal formation amounts to 11×10^{-3} mm/min. It can be seen therefore, that the temperature of maximum devitrification of a specific glass is a most important property from the point of view of working the glass. If during the working or annealing process a glass is held at or near the temperature of maximum devitrification for any appreciable length of time, it crystallizes and ceases to be a glass. In consequence, all glass formulation practice seeks to separate the normal working temperature of the glass from the range of maximum devitrification and it is to this end that many of the various extra components are added to the glass mix. Devitrification of glass can also take place at ordinary temperatures, either under the action of constant vibration, or due to CO_2, SO_2 and other acid gases in the atmosphere, or by a combination of these factors.

4.3 Physical properties of soda glass

The arrangement of atoms in the glass structure is shown in Fig. 4.3.

The softening-point of soda glass, which is considered as corresponding to 10^8 P, is normally given as between 700 and 730°C. The annealing-point is the one at which the viscosity is equal to 10^{13} P and at which stresses in a glass pane or other object, 6 mm thick, can be relieved in 15 min. This temperature is given as about 540°C. The coefficient of thermal expansion of soda glass is between two and three times as high as that of some of the hard glasses such as borosilicate glass and between 10 and 15 times that of fused silica. It is commonly quoted as 8.5×10^{-6}/degC. This high thermal expansion figure means that thick sections of soda glass must never be subjected to differential heating or cooling, as stresses are then set up which will crack the glass. Borosilicate glass, which has a coefficient of thermal expansion of only 3.2×10^{-6}/degC is used for such purposes as oven-ware, glass used in conjunction with furnace construction, etc.

The thermal conductivity of glass is rather high, amounting to 0·045—0·075 W/m.degC, which is of the same order as building brick, concrete, soil and water. The specific heat of soda glass is of the order of 0·93 J/g or about one-fifth of that of water.

The density of glass varies quite considerably according to its formulation. The densities given for soda glasses range from 2·375 kg/dm³ for glass with the composition 80% SiO_2 and 19·5% Na_2O to 2·587 kg/dm³ for glass with 67% SiO_2, 12·7% Na_2O and 18% CaO.

The refractive index of soda glass is one of the lowest of all glasses and amounts to about 1·5. Most grades of optical glasses have refractive indices much higher than this, lead glasses usually averaging around 1·7.

There is a considerable variation in the percentage of light actually transmitted through glass, when thick sections are considered. For good optical glasses, some 63% of light is transmitted through thicknesses of 1 m, which rises to 93% for clear silica. The actual figure for commercial soda glass varies very considerably and is at all times a good deal less. It is markedly affected by such impurities in the glass as manganese and iron oxides. In addition, the transmission of light differs for different wavelengths, longer wavelengths having less penetrating effect.

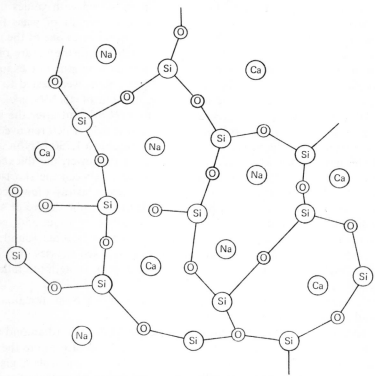

Fig. 4.3   Arrangements of atoms in the structure of glass

Young's modulus for soda lime glasses varies between 57 and 82 GN/m² depending on the composition. The higher value is found with formulations where the percentage of Na₂O is somewhat restricted. Most forms of window glass have a Young's modulus of 76 GN/m², while heavy window glass has 72 GN/m².

When glass is subjected for short periods of time to stresses within the elastic limits, the piece of glass returns to its original shape after the stress has been removed. Unlike other materials, in which the elastic limit is followed by a region of plasticity, glass fails almost immediately it is stressed beyond its elastic limit. Glass is broken by either tension, compression, shear or impact, but in actual fact breakage normally occurs due to rupture only, which is a form of shear failure.

F

The breaking stress R of any glass is given by the following formula:

$$R = (2ET/\pi\sigma c)^{\frac{1}{2}}$$

where E is Young's modulus, T the surface tension, σ Poisson's ratio and c the half-length of the crack. The rupture stress can be assessed to amount to about 20—24 GN/m^2.

The tensile strength of glass fibres is very high indeed with values up to 3,000 MN/m^2. It has been found that the tensile strength of glass fibres increases markedly as the diameter is reduced. This indicates one of the most important aspects of the properties of glass, which is that the nature of the surface of the glass object very markedly affects its strength. For example, certain glass rods which had slight surface imperfections were tested to destruction and were found to have a breaking strength of 45 MN/m^2. After sand-blasting, the strength was reduced to 13·8 MN/m^2. But after the same rods were immersed for a short time in hydrofluoric acid which removed the flaws, the strength of the rods increased nearly 40-fold to 1,730 MN/m^2.

The strength of all glass objects *per unit area* is the lower, the thicker the section. There is also a very considerable difference between the short-term strength of a glass object, when it is subjected to stresses lasting a few seconds and the long-term strength when it is subjected to the same degree of stress for several weeks. The long-term strength is normally only about 35% of the short-term strength. This is one of the reasons why it is often found that when windows are inserted under conditions of unrelieved stresses, they tend to crack some weeks after insertion when the long-term strength has fallen to the critical value.

A safe working stress is normally in the region of 7 MN/m^2 for annealed glass and 28 MN/m^2 for tempered glass.

The hardness of glass, when measured on the Moh scale (diamond=10) varies between 5·4 and 5·8. However, there is some difference too in the way thermal treatment affects the surface hardness of glass. 'Armourplate' glass in which the surface has been specially toughened has a Moh scale figure of almost 9, although once the surface has been penetrated, the body of the glass has the usual figure of around 5·4.

The electrical resistivity of glass varies far more widely than is normally thought. Soda lime glasses have a comparatively low resistivity, and it is found that the actual resistivity is very markedly affected by the percentage of Na_2O in the glass. The following table gives the resistivities of a number of typical glasses:

Glass	Resistivity, $\Omega.cm$
30% Na_2O glass	10^9
20% Na_2O glass	5×10^{10}
10% Na_2O glass	10^{15}
Pyrex glass	3×10^{14}
Lead-borosilicate glass	8×10^{19}

The dielectric strength of glasses varies not only with respect to the chemical composition, but also with regard to the nature of the glass surface. Polished glass can have a strength of 3,000 kV/cm while the value for a coarse ground sample of the same glass is only 500 kV/cm. The dielectric strength per unit thickness also goes up as the thickness of the glass is reduced.

4.4 Chemical resistance of glass

Glass is highly resistant to attack from water and atmospheric pollution as well as dilute solutions of most acids, bases and salts. The standard soda lime glasses have a fairly good resistance to most common reagents but are attacked by concentrated sodium hydroxide solutions, hydrofluoric acid and fluorides, and also by hydrochloric and phosphoric acids to a slight extent. The high lead optical glasses are far more readily attacked by alkaline solutions, while the borosilicate hard glasses have much better resistance. Persistant exposure to polluted atmospheres has the effect of dulling the surface of glass somewhat and increasing the risk of devitrification and consequent loss of strength.

4.5 Surface hardening of glass

To prevent glass from splintering into large sharp pieces, a process called surface hardening is carried out with many glass products, such as plate glass, which is used in areas where a breakage could cause injury to people. The hot plate glass is passed through a zone where cold air is blown over its surface. This contracts the surface structure and compresses the glass underneath, so that it is under continuous stress. Stressed glass is very hard and tough, and difficult to cut. If a fracture is induced in such 'Armourplate' glass, the entire body breaks into small crystals which have quite blunt edges, and minimizes danger to persons. Normal annealed plate glass shatters to form large jagged pieces with razor sharp edges. In nearly all countries the use of surface hardened glass or 'Triplex' safety glass, which is glass sandwiching plastic sheet, for motor cars is obligatory. There are no regulations at present regarding the use of surface-hardened safety glass in the building industry, although numerous accidents take place each year in connections with the shattering of large plate glass windows.

4.6 Coloured and opaque glasses

Such glasses can be produced in three ways:

1 Solution colours are caused by dissolving certain metallic oxides in the melt. The most common ones used are oxides of titanium, vanadium, chromium, manganese, iron, cobalt, nickel and copper. Even small quantities of these impart their characteristic colour to the glass.

2 Colloidal particles impart colour to glass. Ruby glass is formed by the suspension of submicroscopic particles of copper or selenium, while colloidal silver is used for yellow glasses.
3 Certain suspended larger particles give glass a translucent appearance, which can be either white or coloured. Opal glass is produced by particles of titanium dioxide, zirconium oxide or antimony oxide. A coloured opaque appearance is obtained by the use of larger particles of nickel, copper, cobalt, etc., in the form of oxides.

4.7 Attachment of glazed films to surfaces

Glass films can be attached to either:

1 Ceramic surfaces.
2 Metal surfaces.

GLAZING OF CERAMIC SURFACES

Glazes are glasses of low fluidity and are made from complex mixes of various oxides. The following oxides are used, although some of them, which are poisonous, have specialized uses only (see, for example, p. 53):

$$SiO_2 \text{ (always present), } B_2O_3, Al_2O_3, BeO, K_2O, Li_2O,$$
$$CaO, MgO, SrO, BaO, ZnO, PbO, SnO_2 \text{ and } ZrO_2.$$

The appropriate glazing compounds are first of all heated to produce so-called 'fritting'. The ceramic article is then dipped into the glaze bath, the thickness of the deposited glaze depending upon the porosity of the ceramic and also the speed with which the ceramic article is passed through the glaze bath and subsequently withdrawn. It is essential that the coefficient of expansion of the glaze is of the same order of magnitude as that of the ceramic body. If the coefficient of expansion of the glaze is greater than that of the ceramic, the glaze tends to be spalled off. If it is less, it cracks producing fine hair line 'crazes'. Care must also be taken of the maximum devitrification temperature of the glaze used, as the maintenance of the ceramic object at this temperature causes the glaze to be destroyed. When glazes are under compression their tensile strength is increased, as is their cross-breaking strength and thermal shock resistance.

VITREOUS ENAMELLING

Two methods of vitreous enamelling are used. In the wet process a slurry of glass constituents, containing mixed calcium, nickel and aluminium salts, together with ground silica, is painted upon a sand-blasted steel surface at room temperature. The object is heated gently to evaporate the water and is then raised slowly to red heat in order to allow the solids to fuse. A continuous glass is formed on the surface of the steel object which is bonded to the steel surface by ferrous silicate. Coatings applied in this way tend to be rather irregular and somewhat porous.

In the dry process the glass constituents are dusted over the red-hot object. The object is then reheated in order to fuse these glass constituents. The coatings applied in this way vary in thickness between 0·8 and 1·6 mm. This process gives rather better results than the wet one, but costs more to carry out. Vitreous enamelled steel sheeting for cladding purposes in the building industry is becoming more popular. It is essential that pinholes are absent and that such units are not employed under any conditions of mechanical stress or thermal shock. Vitreous enamelling has a superb resistance to chemical agents, but its resistance to shocks of any kind is poor. Once a crack has appeared, even if barely visible to the naked eye, rapid destruction of the steel underneath takes place, due to differential aeration corrosion. This causes the vitreous enamelling to be stripped off, because of the additional stresses produced by the formation of rust underneath the glazed surface. Cracks in vitreous enamelling can be repaired, provided the steel underneath has not started to corrode, by the use of a sodium silicate cement, followed by cold pickling with sulphuric acid which reacts as follows:

$$Na_2SiO_3 + H_2SO_4 \rightarrow SiO_2 \downarrow + Na_2SO_4 + H_2O$$

Vitreous enamelling, being a glass, should not be brought into contact with either strong alkali solutions or with fluorides.

4.8 Glass fibre wool

Glass fibre wool is widely used in the building industry as a thermal insulation material, and is also incorporated into plastics as a strengthening agent (see Fig. 4.4). Glass fibre wool is made by the Crown process, which is somewhat similar to the way 'candy floss' machines work (see Fig. 4.5).

From the forehearth of a glass tank a relatively thick stream of soda glass flows by gravity down a platinum bush into a rapidly rotating alloy dish, which has hundreds of small vents round its periphery. The glass is thrown out through these fine holes by centrifugal force and forms fibres. The thickness of these fibres is controlled by the compressed air which blows them downwards. When the fibreglass wool is to be used for thermal insulation purposes, it is sprayed with phenol formaldehyde resin binder. Afterwards the fibres are subjected to further turbulence to ensure complete random fibre distribution in the mat. The mass is then passed along to curing ovens, which cure the plastic coating, and then to guillotines and trimmers. Glass fibre wool can be produced with fibre thicknesses between 0·001 and 0·006 mm. The building industry tends to employ the coarser grades, the finer grades being mainly used by the aircraft industry.

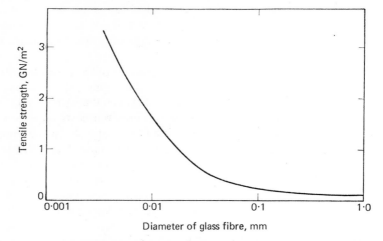

Fig. 4.4   Relationship between the tensile strength of glass fibres and their diameter

Fibreglass mats have densities between 12 and 16 kg/m³ and a thermal conductivity of about 0·04 W/m.degC at ambient temperatures. The material has Class 1 fire resistance when tested according to BSS 476:1953 and can be used at temperatures up to 230°C. It has an angle of capillary contact to water slightly below 90°C and is thus not water repellant, as the plastic foams usually are. When in contact with liquid water, fibreglass material needs protecting with a layer of plastic foil.

The material is, however, not hygroscopic in damp atmospheres nor does it bed down with time.

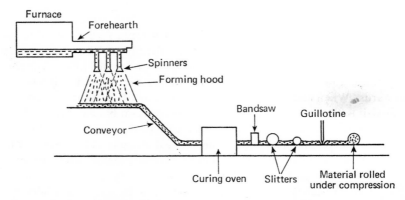

Fig. 4.5   The manufacture of glass fibre (By courtesy of Pilkington Brothers Limited)

MINERAL WOOL

Closely related to glass fibre wool is mineral wool, produced by a similar process but employing a natural mineral called diabase as raw material.

This has the following composition:

$$43\% \text{ SiO}_2, 16 \cdot 5\% \text{ Al}_2\text{O}_3, 11\% \text{ FeO and Fe}_2\text{O}_3, 1 \cdot 5\% \text{ MnO}_2,$$
$$18\% \text{ CaO}, 8\% \text{ MgO and } 2\% \text{ of other materials.}$$

The mineral is melted at a very high temperature in a furnace and then drawn out to thin fibres in a spinning machine. The surface of the glass-like

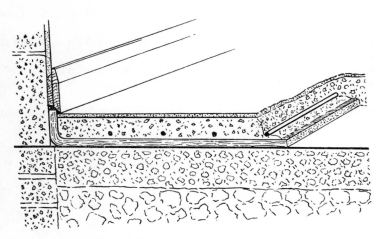

Fig. 4.6   Mineral wool used for underfloor insulation
(By courtesy of Stillite Products Limited)

mineral wool is coated by a mixture of oil and phenolic resin and is thus made water repellant. Like glass fibre wool, the mineral wool can be used as a loose fill (see Fig. 4.6) and its properties are then virtually identical to the former. Mineral wool is also used in the form of slabs, when a combination of insulating properties and strength are required. Slabs of mineral wool are elastic and spring back to their original thickness provided the elastic limit has not been exceeded.

It has been found that steel objects encased in a fibrous material such as mineral wool tend to corrode far faster than they do in air. For this reason it is suggested that such steel objects should either be heavily galvanized or made from stainless alloys. The reason for the rapid corrosion is undoubtedly differential aeration.

Literature Sources and Suggested Further Reading

1. *Glass; The Structure of;* Various authors, 6 volumes, Consultants Bureau, New York (1958–1966)
2. HOLLAND, L., *The Properties of Glass Surfaces*, Chapman and Hall, London (1964)
3. MacKENZIE, J. D., *Modern Aspects of the Vitreous State*, 3 volumes, Butterworth, London (1962)
4. McMILLAN, P. W., *Glass-Ceramics*, Academic Press, London–New York (1964)
5. MOREY, G. W., *Properties of Glass*, Reinhold, New York (1954)
6. PHILLIPS, C. J., *Glass—Its Industrial Applications*, Reinhold, New York (1960)
7. PILKINGTON BROTHERS LIMITED, ST. HELENS, Technical Brochures and Other Information
8. RAWSON, H., *Inorganic Glass-Forming Systems*, Academic Press, London–New York (1967)
9. STANWORTH, J. E., *Physical Properties of Glass*, Clarendon Press, Oxford (1950)
10. STILLITE LIMITED, Technical Brochures
11. VARGIN, V. V., *Technology of Enamels*, MacLaren, London (1967)
12. WEYL, W. A., *Coloured Glasses*, Society of Glass Technologists, Sheffield (1951)
13. WEYL, W. A., and MARBOE, E. C., *The Constitution of Glasses*, 2 volumes, Interscience, New York (1964)

Chapter Five   Wood and Wood Products

In spite of the rapid technological developments which have taken place in the field of plastics, wood is still one of the most widely used materials. It is employed both in the semi-processed state and the fully-processed state such as plywood, chipboard, hardboard, etc.

5.1 Chemical composition of wood

The cell walls of wood, whatever their nature, consist of three main materials which are:

1 Cellulose.
2 Hemicelluloses.
3 Lignin.

In addition, timber contains about 0.2—0.5% protein and between 1 and 5% mineral matter which is left behind as ash when the wood is burned. This mineral matter is normally in the form of Ca^{2+}, Mg^{2+} and K^+ cations and CO_3^{2-}, PO_4^{3-}, SiO_3^{2-} and SO_4^{2-} anions. There are also small quantities of numerous additional components, the nature and quantity of each varying from timber to timber. The most important of these are resins, tannins, terpenes, flavanoids, quinones, glycosides and alkaloids, all of which are rather complicated organic compounds. Wood also contains various types of other carbohydrates, in addition to the cellulose materials mentioned above.

CELLULOSE

All types of wood contain between 40 and 50% of cellulose by weight, which is given the simplified formula: $(C_6H_{10}O_5)_n$, where n is a very large number. The cellulose material consists of long chains of six-membered rings, each ring containing five carbon atoms and one oxygen atom (see Fig. 5.1). Each ring also bears one (CH_2OH) group and two (OH) groups, being joined to the next ring by an ether $(-O-)$ bond. The number of basic rings in an average

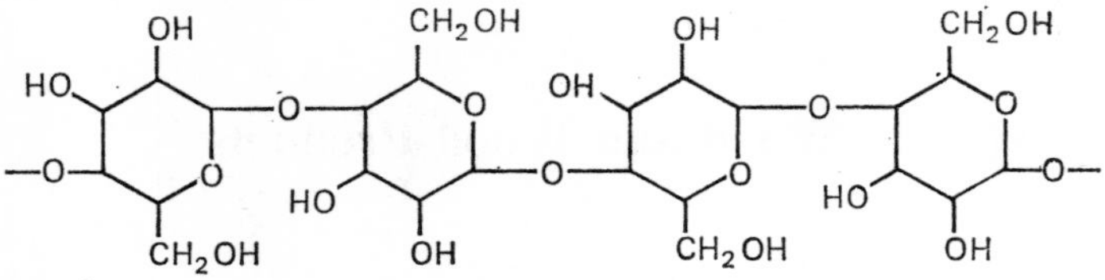

Fig. 5.1 The structure of cellulose

cellulose molecule is about 10,000. Because of their chain nature, cellulose is able to form fibres, similar to many of the other natural vegetable fibres such as cotton, hemp, jute, etc. Cellulose can exist in a regular crystalline arrangement or in an amorphous form, or both.

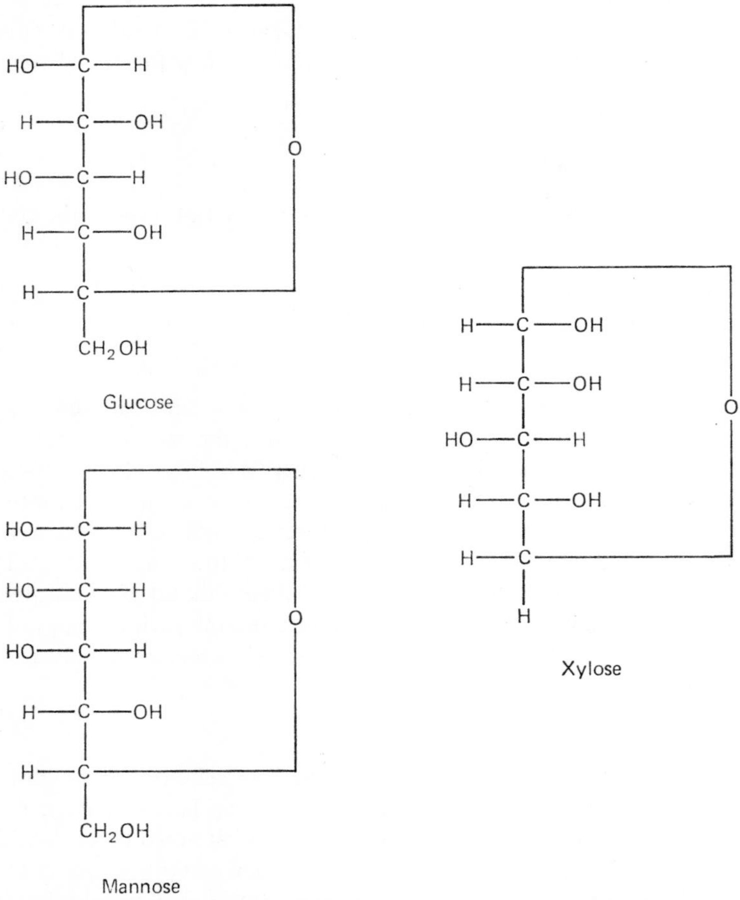

Glucose

Xylose

Mannose

Fig. 5.2 The structure of three typical sugars present in hemicellulose

HEMICELLULOSES

These are materials with the same general formula as normal cellulose, but with a very different constitution. While cellulose consists only of glucose units, hemicellulose is made up of a variety of different sugar units (see Fig. 5.2). In addition, the chains are much shorter so that the material is not fibrous like cellulose but more gelatinous: hemicellulose reacts much more readily than cellulose with either dilute acids or dilute bases.

LIGNIN

Lignin is an extremely insoluble substance, and like the other two materials mentioned, is a polymeric material. The basic constituent of the lignin polymer is a phenyl propane nucleus connected to other chemicals such as vanillin and

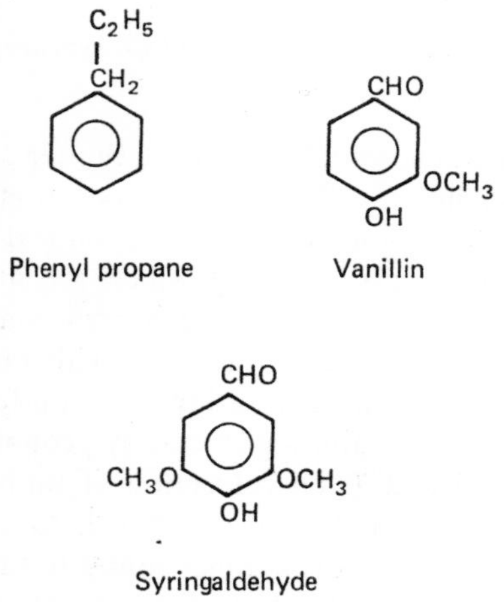

Phenyl propane   Vanillin

Syringaldehyde

Fig. 5.3 The structures of three of the main constituents of lignin macromolecules

syringaldehyde to form a highly complex branched-chain structure (see Fig. 5.3). Lignin is resistant to hydrolysis by acids, but can be broken down by sulphonation processes and treatment with alkalis at high temperatures. It is also dissociated by chlorination followed by reaction with alkalis which dissolves the chlorolignin.

In general, softwood contains about 27% lignin, while hardwood has 21%. Hardwood usually contains about 20% pentosans, which are polysaccharides built up from pentose (five carbon atoms) sugar residues, while the average percentage of pentosan in softwood is only about 8—9%.

5.2 Colour of wood

The main structural components of wood, which are cellulose, hemicelluloses and lignins, are virtually colourless. The colour of wood, which ranges from almost white down to ebony black, is produced by natural colouring matter in the timber, which is mainly flavones, quinones and related materials. Some of these colouring materials are part of the polymeric substances which make up the timber so that they are not readily extracted by solvents. However, many of the natural dyes are readily destroyed by strong sunlight. This can be prevented by incorporating an ultraviolet-absorbing substance in the lacquer applied to the wood. Materials often used for this purpose are substituted benzophenones such as 2,2',4',4,tetrahydroxybenzophenone (Fig. 5.4).

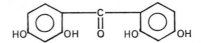

Fig. 5.4   The structure of 2,2',4,4'-tetrahydroxybenzophenone

Some timbers contain water soluble colouring matter and timbers such as afzelia, ayan and idigbo tend to suffer from discoloration when subjected to dampness. Timbers which contain free tannin are stained when brought into contact with iron and iron salts. Timbers which are affected in this way include: oak, sweet chestnut, walnut, makore, afrormosia and idigbo. When timber of this kind is to be used in conjunction with metal fastenings it is necessary to specify either galvanized fittings or non-ferrous metals. The staining of such timbers by iron salts is particularly pronounced when the pH of the timber is between 5 and 7. Discoloration of timber by contact with iron is also found with Western red cedar and North American yellow cedar. These are iron complexes with the tropolones found in these timbers.

Timbers that contain tannin are also darkened under the action of ammonia, which is often liberated when some types of glues deteriorate. Both alkalis and acids tend to stain timber. Alkalis turn timber brown or yellowish-brown, while acids stain such woods as walnut, sycamore, maple, agba and sapele pink because they contain materials called leucoanthocyanidins.

5.3   General properties of timber

Because wood is a biological material, it shows a wide variation in physical properties, both between different botanical species and even within the same species. All strength characteristics vary enormously in the three growth directions, i.e. longitudinal, tangential and radial, with strength being far higher (often by a factor of 10 or more) in the longitudinal direction. Densities of wood vary between 0·04 kg/dm³ for some of the balsa woods, to 1·24 kg/dm³ for *lignum vitae*.

One of the main adverse properties of timber is its liability to moisture absorption when in contact with liquid water or under conditions of high humidity (see Fig. 5.5), and also its tendency of shrinking when kept in a dry atmosphere. Apart from the swelling and shrinkage of the timber, which is the main effect, distortion takes place as the degree of absorption is not uniform. Because of internal stresses which arise, splitting and disintegration of the timber can take place under extreme conditions.

Water is held by the cell structure of the wood as water of constitution, as surface bound water and as capillary condensed water. Water of constitution has reacted with some of the polymers in the wood. The fibre saturation point of most timbers is about 30% and is defined as the moisture content that exists when wood is in equilibrium with air with 100% humidity.

Shrinkage varies enormously with the direction of the timber. It is greatest in a tangential direction and least in a longitudinal direction.

Under any condition of swelling and expansion, the ratios applicable in the different directions are as follows:

$$\text{Longitudinal direction: } x$$
$$\text{Radial direction: } (50\text{—}60)x$$
$$\text{Tangential direction: } (100\text{—}120)x$$

The timbers with the greatest tangential movement are the following: gurjon, European beech, European oak, sycamore and ramin, among the hardwoods, which move up to 3% when shifted from an atmosphere containing 90% relative humidity, to one containing only 60% relative humidity. Western hemlock and Scots pine are two softwoods with tangential movement of about 2%. The lowest tangential movement among the hardwoods is found with teak, idigbo, iroko and mahogany among the hardwoods, and Douglas fir among the softwoods, which show a movement of only about 1% under the conditions mentioned before.

When green wood is dried, the strength of the timber remains constant until the fibre saturation point is reached. After that, strength increases linearly with further reduction in moisture content.

As the angle of capillary contact of timber is already very high, with figures of 90° and more having been reported, the water absorption of timber can, in general, not be reduced simply by impregnating the material with a substance designed to raise the angle of capillary contact still further. Any treatment of

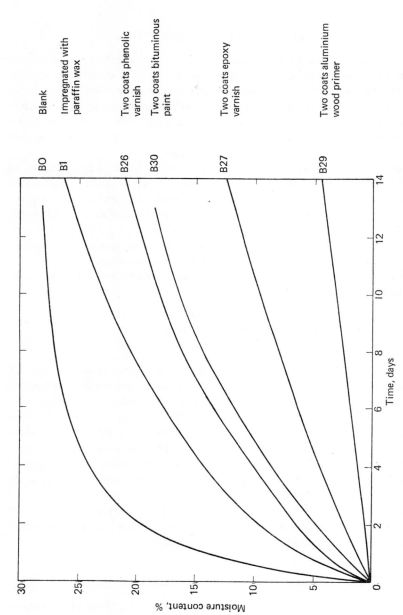

BO — Blank

B1 — Impregnated with paraffin wax

B26 — Two coats phenolic varnish

B30 — Two coats bituminous paint

B27 — Two coats epoxy varnish

B29 — Two coats aluminium wood primer

Fig. 5.5 Moisture uptake by beech heartwood in a wet atmosphere after treatment with various moisture retardants (By courtesy of the Timber Research and Development Association)

timber to prevent moisture absorption is mainly designed to reduce the rate of moisture ingress. No timber treatment can in any way affect the final equilibrium moisture concentration in the timber, but the achievement of equilibrium can be delayed appreciably. With small samples such a delay may be a matter of weeks, while with larger components the delay may amount to several months. As climatic conditions usually vary appreciably, the effect of protective treatment is to flatten the moisture absorption curve. This provides considerable protection against warping and splitting.

Impregnation with such materials as paraffin wax, chlorinated naphthalene wax, polyester resin and epoxy resin reduces the impregnation figure a good deal, but such methods are expensive to carry out and are in most cases inferior to the application of surface coatings.

The best moisture retardants are short oil varnish, epoxy resin varnish and chlorinated rubber modified varnish. Long oil phenolic resin varnishes give good protection against water vapour, but not against liquid water. An aluminium wood primer made by dispersing exfoliated aluminium in an epoxy resin varnish is by far the best moisture retardant used in the building industry. It reduces water absorption in 24 hr to about 2%. It is superior to a coating of bitumen more than three times as thick.

Finally, a method used of making types of wood/plastic components which have virtually zero moisture absorption is to pressure-impregnate dried wood with epoxy resin, followed by curing with high-energy gamma rays. Such resin-impregnated wood, which has up to 30% resin content, is virtually immune from swelling, is resistant to insect and vermin attack, yet retains the pleasing appearance of timber. Its uses are mainly the manufacture of door handles and other fittings which are exposed to the weather. A large number of other resin-impregnated types of wood are also made, usually utilizing phenol-formaldehyde and urea-formaldehyde resins.

HEAT TREATMENT AND ACETYLATION

Two methods of reducing the liability of wood to swell are heat-treatment and acetylation.

Heat-treatment

When timber is heated in the absence of air, usually underneath a molten metal, it becomes resistant to swelling and shrinking. However, its colour darkens and its strength is reduced. Toughness reduces by about 40% whereas hardness and modulus of rupture in bending are reduced by 20%.

Acetylation

One of the reasons why wood is so hygroscopic is because of the existence of numerous hydroxyl groups in the structure. By reacting these hydroxyl groups with acetic anhydride, non-polar groups are formed which have a far lower affinity for water. Acetylation of the wood is carried out in the vapour phase by using a mixture of acetic anhydride and pyridine. Softwoods need to be acetylated to a greater extent than hardwood to obtain the same degree of anti-swelling efficiency. Acetylation normally reduces the liability of wood to shrink or swell by about 70%. Acetylated wood looks almost like normal wood and its strength appears to be unaffected by the acetylation treatment. It has a very much increased resistance to all types of fungus and insect attack.

CHEMICAL RESISTANCE OF TIMBER

Timber has good resistance to polluted atmospheres and is particularly useful in its resistance to acid fumes, which attack steel. Constructional softwoods, such as Douglas fir, pitch pine, larch and redwood, are the best in this respect, as is the hardwood teak.

Timber is unattacked by non-oxidizing mineral acids, if the concentration is less than about 25%, and by most organic acids with a concentration below 60%. Timbers with a high cellulose and lignin content are the best in this respect. Other liquids which are harmless to timber are: aluminium, ammonium, iron and copper salt solutions, alcoholic solutions, hydrocarbons, formaldehyde, cement slurries and brine.

Timber is attacked by solutions of nitric acid, nitrates, chlorates, alkalis, phenol, calcium and zinc salts, as well as salts of strong bases and weak acids such as sodium sulphide and sodium carbonate. Also, timber surfaces can be corroded away in time by sulphuric acid solutions. Even such weak alkaline solutions as detergents and soaps will eventually soften a timber surface.

5.4 Causes of deterioration of timber

The main causes for the deterioration of timber are the following:

1 Mechanical wear.
2 Chemical action.
3 Weathering.
4 Infection by fungi.
5 Infestation by insects.

Mechanical wear of timber naturally depends upon the surface hardness of the material and the extent to which it is exposed to driven sand and other abrasive agents. In general, such wear is of little practical importance.

Strong acids and bases may decompose certain constituents of timber and thereby cause disintegration of the structure. If timber is kept in contact with such materials, it must be suitably protected by plastic cladding materials.

Weathering is mainly the result of alternating shrinkage and swelling of the surface of the wood, combined with the action of sunlight and atmospheric pollution. The normal rate of weathering is about 1 mm every twelve years.

Even this slow rate of surface deterioration can be prevented by adequate painting.

The most important destructive agents are attack by fungi and insects.

FUNGAL ATTACK

There are four main types of fungi (the life of which is given in Fig. 5.6) which attack timber. These are:

1 Dry rot (*Merulius lacrymans*).
2 Cellar fungus (*Coniophora cerebella*).
3 Wet rot (*Poria vaillantii*).
4 Greenhouse rot (*Poria xantha*).

Of these the most common fungus which attacks timbers in buildings is dry rot. Wood which has been decayed by this splits longitudinally and across the grain into cubes which are up to 5 cm in size. The decayed wood is brown and

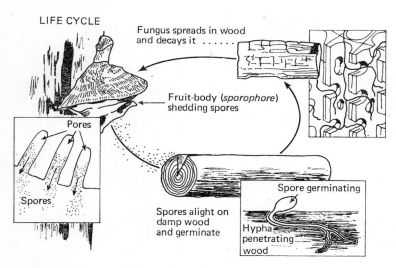

Fig. 5.6   Life cycle of a fungus
(By courtesy of the Timber Research and Development Association)

crumbles into a powder when touched. To enable dry rot to start, the timber has to have a moisture content of at least 20%, and probably between 30 and 40%. The first stage is a cotton wool texture film, which then changes to long strands of white–grey matter, which is able to bypass up to 1—1½ m of brickwork, to infect portions of timber beyond.

Cellar rot requires really damp conditions and is found in timber containing between 40—50% moisture. The fungus, dark brown in colour, is also in the form of strands, which however, do not become as long as the dry rot variety.

G

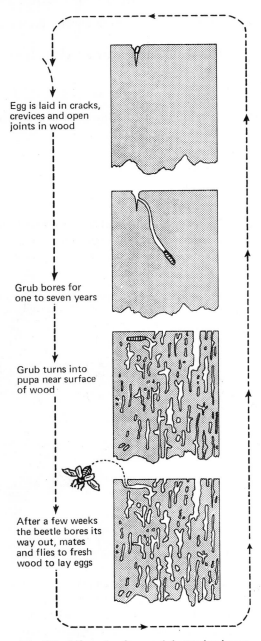

Egg is laid in cracks,
crevices and open
joints in wood

Grub bores for
one to seven years

Grub turns into
pupa near surface
of wood

After a few weeks
the beetle bores its
way out, mates
and flies to fresh
wood to lay eggs

Fig. 5.7   Life cycle of a wood-destroying insect
(By courtesy of the Timber Research and Development Association)

Wet rot is mainly found externally in waterlogged timber such as posts, fencings, etc., which has not been adequately surface-protected by creosoting or other treatment. This form of rotting is unlikely to occur with timber which has a moisture content below 40%. The greenhouse rot, as the name implies, is found in very moist and warm atmospheres as in greenhouses, hothouses, etc.

In addition there are a number of fungi which do not attack timber in depth but only stain or soften the surface. One of these is soft rot, which affects submerged timbers. Softwood is, in general, more resistant than hardwood to this kind of attack.

INSECTS WHICH ATTACK TIMBER

1 *Common furniture beetle* (*Anobia punctatum*). This attacks structural and other timber, which is at least ten years old, but cases of attack on plywood younger than this have occurred. The larvae attack the sapwoods of most hardwoods and softwoods, as well as the heartwood of beech, birch, sycamore and elm.

2 *Death-watch beetle* (*Xestobium rufovillosium*) only attack timber which has already been attacked by fungus. It mainly affects hardwood.

3 *Lyctus beetles* (*Lyctus brunneus* and *L. linearis*) affect timber which contains free starch such as oak, ash, elm, sweet chestnut and some tropical hardwoods.

There are a number of other wood-boring insects, such as house longhorn beetles (*Hylotrupes bajulus*), which are fairly rare, and the pin hole borers or ambrosia beetles, which are common only in tropical and subtropical areas, as are the termites, which are probably the most destructive of all. (There is a statutory requirement laid on owners of property infested by house longhorn beetles to report the occurrence to the Local Authority.)

A diagram of the life-cycle of a wood-destroying insect is given in Fig. 5.7.

5.5 Types of timber preservative used

Timber preservatives usually fulfil a dual role, i.e. they protect the timber both against fungus and against insects. The main timber preservatives in common use are the following.

Tar oil and creosote

These are by-products of the dry distillation of coal or timber, and are a composite mix of hydrocarbon oils, phenols, cresols and many other materials. They are resistant to leaching and are therefore particularly useful for treatment of timber used externally. Creosoting very much delays moisture movement in timber and the chemicals contained in it are toxic to fungi, insects and marine borers. Creosote is not corrosive to metals in contact with it, but has the following disadvantages:

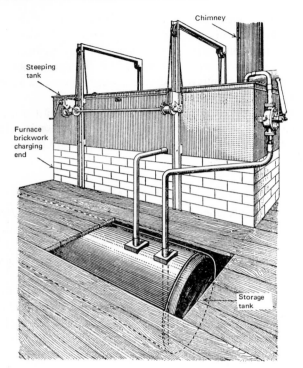

Fig. 5.8   Timber impregnation equipment
(By courtesy of the Timber Research and Development Association)

1 The creosote has a strong odour, which can be picked up by foodstuffs.
2 Creosote wood is difficult to paint over.
3 Plaster and brickwork in contact with it are often stained.

Once the volatile fractions of the creosote have evaporated, timber impregnated with creosote is no more inflammable than untreated wood.

Paraffin oils with dissolved agents
A cheaper substitute for creosote is the impregnation of timber with paraffin oils which contain pentachlorophenol and tetrachlorophenol. Sump oil is often used as a carrier.

Waterborne preservatives (inorganic)
Several metal salts are used in the form of aqueous solutions to impregnate timber. These have various advantages over creosote. Being colourless they do not stain and are usually both odourless and non-flammable. They are cheaper to buy and to transport than creosote but some are liable to be leached out of the timber when exposed to wet conditions. Application is

more difficult than with creosote, but the treated and dried timber can be readily painted. The main types of materials used are the following:

1 *Copper – chrome salts, with or without traces of arsenic*. These consist of mixtures of potassium dichromate, copper sulphate and chromic acetate. The timber is first soaked in the dichromate which penetrates the pores. When this treated timber is then immersed in the copper sulphate and chromic acetate solutions, an insoluble compound is deposited in the structure, which is quite inocuous to foodstuffs, but highly toxic to both fungi and insects.

2 *Tanalith process*. This is an insoluble chromate – arsenate complex deposited in the pores of timber. It is formed by mixing fluorides, arsenates, chromates and dinitrophenol. If monoammonium phosphate is added in addition to the above, the timber is also reasonably well protected against fire risks.

3 *The BM process* consists of impregnating the timber with a mixture of zinc chloride and aluminium sulphate. This protects the timber, not only against fungal and insect attack, but also to a limited extent against fire risks.

4 *Copper – chrome – zinc chloride process*. This consists of 7% cupric chloride, 73% zinc chloride and 20% sodium chromate. It has good preservative properties and excellent leaching resistance.

Numerous other combinations of copper, arsenic, zinc, chromium and mercury salts, as well as silicofluorides and sodium fluoride, are used throughout the world for timber preservation.

ORGANIC TIMBER PRESERVATIVES

These have the advantage of being resistant to leaching, do not stain and can be readily painted over. They tend to penetrate timber more readily than other types of preservatives and can be applied simply by brushing, spraying or cold dipping. On the other hand the volatile solvents used are inflammable, although the final timber does not show any inflammability greater than untreated timber.

The most common organic preservatives used are:

1 Metallic naphthenates.
2 Pentachlorophenol (C_6Cl_5OH).
3 Copper pentachlorophenate ($C_6Cl_5O)_2$ Cu.
4 Benzene hexachloride (C_6Cl_6).
5 Dieldrin.
6 Pentachlorophenyl laurate ($C_{12}H_{23}COO \cdot Cl_5C_6$).

All these materials are toxic to fungi and insects.

5.6 Fire protection of timber

Although wood is an organic material and as such combustible, the speed of burning of wood is slow due to the fact that it is a poor conductor of heat (see Fig. 5.9). In consequence its performance from the fire-resistance point of view in buildings sometimes compares very favourably with that of steel and aluminium members, which often buckle and melt before an equivalent timber beam has burned through.

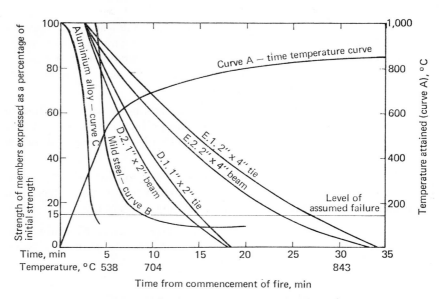

Fig. 5.9 Comparison of performance of timber and metal beams in a fire
(By courtesy of the Timber Research and Development Association)

When wood is heated, the temperature rises steadily up to 100°C. At this point there is a check as the moisture in the timber is evaporated. After this has taken place, the temperature rises again. Decomposition starts at 250 to 300°C with the formation of charcoal and inflammable gases such as CO and CH_4. Such gases are the cause for the flaming of wood.

The final stage of the burning is the combustion of the charcoal, which results in smouldering with the ultimate formation of ash.

Of the total available heat of wood which is 20 MJ/kg, about half to two-thirds is liberated by flaming and the rest by smouldering. Flames rise vertically so that timber in such a position is a greater fire hazard than horizontal timbers such as floor or beams.

The fire-resistance of individual timber varies widely. The highest fire-resistance is shown by greenheart, Indian laurel, teak and Burma padauk.

Then come such timbers as oak, ash, beech, Douglas fir, idigbo and sweet chestnut. Medium fire-resistance is shown by birch, elm, mahogany, maple, Scots pine, larch and nargusta.

Low fire-resistance is shown by cherry, sequoya, walnut, horse chestnut, spruce and most pines. Very low fire-resistances are obtained with alder, Western red cedar, lime, poplar and willow, while balsa wood, as might be expected, has the lowest fire-resistance of all. In general, the denser timbers burn more slowly but the presence of resins, gums and mineral matter has a considerable influence on the rate of burning as well.

Tests have been carried out to determine the time required for the strength of typical softwood timber beams to be reduced to 15% of their original values, when subjected to temperatures of up to 850°C, and to compare the figures with equivalent metal beams. The results were as follows:

Beam	Time, min
1 in × 2 in beam (2·5 cm × 5 cm)	13·5
2 in × 4 in beam (5 cm × 10 cm)	24·5
Aluminium alloy beam	4·0
Mild steel beam	9·5

METHODS OF IMPROVING THE FIRE-RESISTANCE OF TIMBER

Impregnation of timber with solutions of chemical salts
Most of the fire-retardant salts used work in the following way. They decompose at a temperature somewhat below the decomposition temperature of the timber, and produce non-flammable gases which mix with the highly inflammable gases given off by the timber. This forms a protective blanket against the attacking flame and inhibits flame propagation at the surface of the material. At the same time the decomposition of the fire-retardant is an endothermic process, which cools the igniting flame. Often the fire-retardant reduces the rate of burning by forming a melt at a temperature below the ignition point of the timber. The glaze formed protects the timber against combustion. Ammonium phosphate induces the formation of heavy and dense layers of charcoal, which do not smoulder very rapidly. The most widely used salts for impregnating timber against fire are the following:

1 *Ammonium salts* such as phosphates, sulphates, chlorides, etc., are employed. Of these, sulphates and chlorides are corrosive to metals, and can be leached out in water. Timber also becomes more hygroscopic when such salts are used.

2 *Boric acid.* This is not hygroscopic and is also not leached out of the timber. On the other hand, the material is not as effective in the prevention of flaming as the ammonium salts.

Both types of flame retardants are usually applied by pressure impregnation.

Flame-resistant surface coatings

The paints used for this purpose usually contain either sodium silicate, potassium silicate or calcium sulphate with an inert filler, such as asbestine or china clay. They are applied thickly by spray or brush and it is essential that they be put on to a clean timber surface. In the case of a fire, the sodium silicate swells up, to form a frothy mass, which protects the wood against heat. Neither sodium silicate paints nor those based on calcium sulphate are stable against water or high humidity conditions. A more modern type of coating is one using a combination of urea-formaldehyde resin and ammonium phosphate. This also forms a spongy insulating layer on the surface of timber during a fire, but unlike the silicates, it is unaffected by moisture.

Fire-resistant paints are easily applied and are cheaper than the impregnation techniques. The protection afforded is, however, not as good.

5.7 Plywood

Most plywood used in the construction industry is made from either Douglas fir, Western hemlock, larch, spruce or soft pines. Veneers are cut off the surface of logs and are glued together so that the grains cross over each other. After the glue has been spread and the veneers assembled, the panels are cured at a pressure of about 10 bar. Two basic grades of plywood are made:

1 *Exterior grade* plywood which employs a hot-pressed phenolic resin glue as binder. This glue is virtually insoluble under any exposure conditions, including boiling water.

2 *Interior grade*, which accounts for some two-thirds of the production, uses protein glues which have a soya bean–blood base. Hot pressing is generally used. While interior grade plywood is capable of withstanding temporary wetting, it is completly unsuitable for conditions where there is continuing dampness.

Each grade is made in various standards, depending upon the use envisaged for it. The N-grade is the best, followed by grades A, B, C and D (the lowest).

MECHANICAL PROPERTIES OF PLYWOOD

The following expression has been obtained:

$$e = 1 \cdot 075 \times 10^{-6} \left(\frac{P}{1-P} \right) + 3 \cdot 2 \times 10^{-6}$$

where e (cm/cm.degC) is the coefficient of expansion and P the fraction of panel thickness in plies (veneers) with grain perpendicular to the direction of expansion. The average coefficient of thermal expansion is equal to:

$$2 \cdot 79 \times 10^{-5} \text{ cm/cm.degC}$$

The *water vapour permeability* of plywood is as follows:

1 Exterior grade plywood without surface finish:

$\frac{3}{8}$ in (9·1 mm) $1·45 \times 10^{-7}$ sec/m

$\frac{3}{8}$ in (9·1 mm) overlaid plywood $0·6 \times 10^{-7}$ sec/m

2 Interior grade plywood with undercoating and shellac finish:

$\frac{1}{4}$ in (6·3 mm) $1·5 \times 10^{-7}$ sec/m

In contrast, a concrete block, 8-in (20 cm) thick has a water vapour permeability of $4·8 \times 10^{-7}$ sec/m.

Table 5.1 gives the main mechanical properties of exterior grade A plywood.

TABLE 5.1

	Direction of stress to face grain		
Ultimate strength, MN/m² under	Parallel	Parallel or perpendicular	Perpendicular
Tension	15·2		13
Compression	11·2		9·5
Bending	15·0		13·0
Rolling shear in plane of plies		0·55	
Rolling shear in perpendicular to plies		1·8	
Modulus of elasticity in bending, GN/m²		11	

5.8 Miscellaneous timber products

INSULATION BOARD

Insulation board is a fibre board, which has a density below 0·4 kg/dm³. In its manufacture, greenwood timber is first converted into pulp by grinding and mixed with certain other materials such as sugar cane fibres, repulped waste paper, etc., to give the desired properties to the finished board. Defibration of the timber particles is usually carried out by the Asplund process, in which the particles are heated for 1 min with steam at 10 bar (gauge) which softens the lignin. This enables the fibres to be separated in an attrition mill. The pulp is then refined, sized by the addition of small quantities of phenol-formaldehyde resins, etc., and a suspension of the pulp is allowed to flow on to a closely woven gauze. Water is withdrawn by drainage through this gauze, the mat is passed through rollers and then through a tunnel drier where it meets a fast-moving stream of very hot air. This produces a rigid fibre board with an open structure and low density, which is extremely useful for insulation purposes.

HARDBOARD

This is a very much harder type of fibre board, and its density usually ranges between 0·8 and 1·2 kg/dm³. The production of a fibrous pulp is similar to that used for the manufacture of insulation board. A small quantity of wax emulsion is added to the pulp to act as a water-proofing agent, followed by some 2—3 % of phenol-formaldehyde resin. This greatly improves the strength of the final hardboard. The fibrous slurry is then placed onto the gauze belt and extraneous water is driven off. The wet sheets, which still contain between 65 and 70% of water, are passed through a press at a temperature of 180—210°C and are compressed at about 48 bar. Under these conditions the water in the fibrous mat is both pressed out and evaporated. During the pressing process the mat is supported by the wire gauze in the press. This is necessary as otherwise the hardboard breaks up during drying. Finally, the finished hardboard is heated for several hours at about 160°C, which increases the strength of the board.

Water-resistant hardboard is made by impregnating the board with linseed oil or tung oil. The hardened board is dipped into a bath containing the oil and then exposed to a temperature of 160°C for 10 hr. This polymerizes the drying oil within the fibres of the board.

PARTICLE BOARD

Particle board or chipboard is made from solid fragments of wood which are held together to form a rigid mass by the use of a synthetic resin binder. Most types of chipboard have densities between 0·4 and 0·8 kg/dm³. The chips are first of all dried to a moisture content of 5—12% and are mixed with either urea-formaldehyde or phenol-formaldehyde resin. Urea-formaldehyde is cheaper and has a lighter colour and is preferred for indoor use, while phenolic resins are more durable. The resin is applied to the particles by spraying into a sealed drum in which the chips are being agitated. Pressing is carried out either by extrusion or flat pressing at a temperature of between 100 and 140°C. In addition to single-layer board, three-layer board is made in which the central core consists of coarse irregular particles, while the two external layers are from large flat particles.

Literature Sources and Suggested Further Reading

1. Brauns, F. E., and Brauns, D. A., *The Chemistry of Lignin*, Academic Press, New York (1960)
2. Browning, B. L., *The Chemistry of Wood*, Interscience, New York (1963)
3. American Plywood Association, Tacoma, Washington, Technical Brochures and Information (1963–1968)
4. Desch, H. E., *Timber*, MacMillan, London (1968)
5. Findley, W. P. K., *The Preservation of Timber*, Black, London, (1962)
6. Findley, W. P. K., *Dry Rot and Other Timber Troubles*, Hutchinson, London (1953)
7. Farmer, R. H., *Chemistry in the Utilization of Wood*, Pergamon Press, Oxford (1967)

8. HILLIS, W. E., *Wood Extractives*, Academic Press, New York (1962)

9. PANSHIN, A. J., *Textbook of Wood Technology*, 2 volumes, McGraw-Hill, New York (1964)

10. PERKINS, N. S., *Plywood*, American Plywood Association, Tacoma, Washington (1962)

11. WOOD, A. D., *Plywoods of the World*, Johnston and Bacon, Edinburgh (1963)

12. WOOD, A. D., and LINN, T. G., *Plywoods*, Johnston and Bacon, Edinburgh (1950)

13. TIMBER RESEARCH AND DEVELOPMENT ASSOCIATION, Technical Publications and Information. *Address:* The Building Centre, Store Street, London WC1

Chapter Six   Aluminium, Steels and Alloy Steels

The most common structural metals used are, on the one hand, aluminium and its alloys, and on the other, the many grades of steels and alloy steels, which serve both traditional and novel requirements.

6.1 Aluminium

METALLURGY

Aluminium is the third most widely distributed element in the earth's surface, being surpassed only by oxygen and silicon in this respect. Unfortunately it is mostly present in the form of complex silicates such as feldspar $NaAlSi_3O_8$ and clay $H_4Al_2Si_2O_9$. Up to the present day no economical methods have been developed for extracting aluminium from these compounds.

Aluminium production (Fig. 6.1) is therefore based upon the mineral bauxite, $Al_2O_3.2H_2O$, which is found in large deposits mixed with quantities of silica and iron impurities. In order to make economic extraction of aluminium possible, the bauxite should contain at least 55% Al_2O_3.

Bauxite is first powdered and calcined in order to oxidize any organic matter present. After this it is ground with sodium hydroxide solution of specific gravity 1·45 and treated with steam at about 5 bar (gauge). Any silica present forms an insoluble precipitate of $Na_2O.Al_2O_3.SiO_2.9H_2O$. This is then filtered off from the solution of sodium aluminate $NaAlO_2$ by using filter presses, finally passing the solution through wood pulp in lead-lined vats. The sodium aluminate is diluted and small quantities of alumina are added for seeding purposes. The pure aluminium hydroxide is left to sink to the bottom of the vessel, filtered off, roasted and electrolysed in an electric furnace mixed with cryolite (Na_3AlF_6). The furnaces are operated at between 30,000 and 40,000 A and are kept at a temperature of 1,000°C. Carbon anodes are used and burn to CO_2 during the reaction.

The reactions which take place are summarized as:

$$Al_2O_3 + 2NaOH \rightarrow 2NaAlO_2 + H_2O + \text{solid impurities}$$
(impure)   (soln)

$$2NaAlO_2 + 4H_2O \rightarrow 2Al(OH)_3 + 2NaOH$$

$$2Al(OH)_3 \rightarrow Al_2O_3 + 3H_2O$$
(pure)

$$2Al_2O_3 + 3C \rightarrow 4Al + 3CO_2$$

The aluminium is tapped off from the bottom of the furnace. The last reaction given is highly endothermic with $\Delta H = +3,044$ kJ, which means that the process is very power hungry. For this reason aluminium has frequently been referred to as 'solid electricity' and can only be manufactured economically with the use of cheap electricity such as off-peak hydroelectric or nuclear

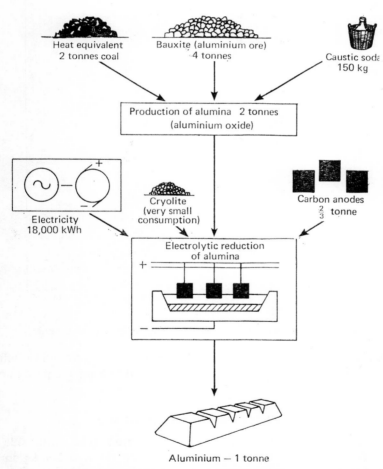

Fig. 6.1   Principal materials and stages in the production of aluminium
(By courtesy of the Aluminium Federation)

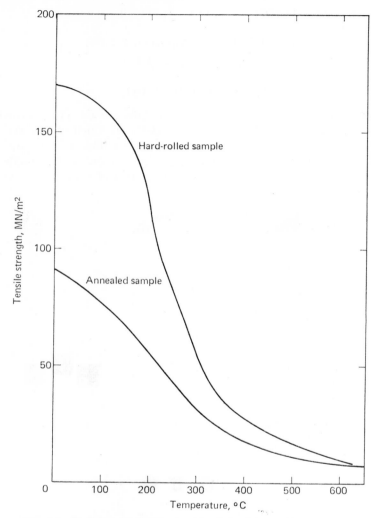

Fig. 6.2 Variation of tensile strength of aluminium with temperature

power. The aluminium normally produced is about 99·8 % pure, which suffices for most purposes. Electrolytic refining can be used to bring the purity up to 99·99 %.

PROPERTIES OF ALUMINIUM

The crystal form of aluminium is a face-centred cube, with side of $4·045 \times 10^{-10}$ m. The density of aluminium is usually taken as 2·70 kg/dm³ at 20°C but lower values than this have been found for cast metal, which are explained by the porosity of such forms of aluminium. The specific heat of

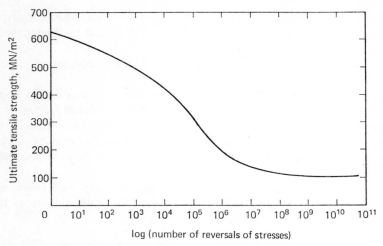

Fig. 6.3   Fatigue curve of high-strength aluminium alloy 75S-T$_6$

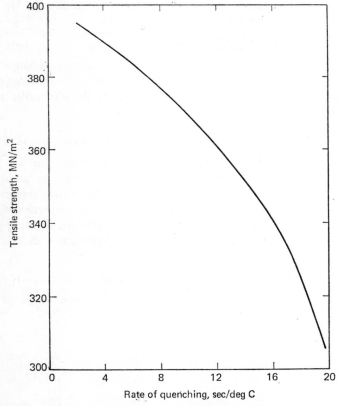

Fig. 6.4   Effect of rate of quenching on the tensile strength of aluminium

aluminium is equal to 0·925 kJ/kg or 22·6% that of water. The melting-point of normal commercial aluminium is given as 659°C and the linear thermal expansion is equal to $2·40 \times 10^{-5}$ degC^{-1}.

The thermal conductivity of aluminium is high, amounting to 218·5 W/m.degC at 20°C.

The electrical resistance of aluminium is roughly 1·5 times that of copper, amounting to $2·845 \times 10^{-6}$ Ω.cm. As, however, its density is only 2·70 kg/dm^3 as against the density of copper of 8·9 kg/dm^3 the conductivity of aluminium per unit weight is actually twice as high as that of copper. Hence the popularity of aluminium cables with steel cores for long distance power transmission.

In its radiating power or emissivity, aluminium ranks well below many other materials. If one takes the emissivity power of a black body as 100% the values which apply to aluminium are as follows:

New aluminium foil (smooth)	4%
New aluminium (patterned surface)	5·7%
Aluminium foil suspended vertically for 3 years	5·0%
Aluminium foil used as a roof panel for 8 years (dusty surface)	8·3%
Aluminium foil after exposure for 2 years to sea air	10%
Aluminium foil chemically oxidized by treatment in hot solution of sodium carbonate and sodium chromate	9·0%

For this reason aluminium foil is widely used as a radiation resistant surface for thermal insulation against heat and cold. Reflectivity is much higher in the visible part of the spectrum than in the infrared. In the ultraviolet region it rises to 90%.

MECHANICAL PROPERTIES OF ALUMINIUM AND ITS ALLOYS

These vary very widely between aluminium of different purities and are markedly affected by small traces of alloying elements. In addition the properties are also markedly affected by the manner of heat treatment and temper prior to the test (see Figs. 6.2, 6.3 and 6.4). The properties of most of the common forms of aluminium and its alloys used in the building industry are given in Table 6.1, which is produced here by permission of the Aluminium Federation.

TABLE 6.1   PHYSICAL PROPERTIES OF ALUMINIUM AND ITS ALLOYS

BS number	Composition	State	Tensile strength, MN/m²	Elongation on 5 cm, %
A. Sheet and strip to BS 1470				
S1	99·99% Al, Cu, Si, Fe 0·1%	0*	62	45
		Hard	102	6
S1C	Min. 90% Al, 0·1% Cu, 0·5% Si,	0	100	30
	0·7% Fe, 0·1% Mn, 0·1% Zn	Hard	140	3
NS3	0·1% Cu, 0·6% Si, 0·7% Fe,	0	115	30
	1·2% Mn, 0·2% Zn, 0·2% Ti, remainder Al	Hard	1/8	3

TABLE 6.1—*continued*

BS number	Composition	State	Tensile strength, MN/m²	Elongation on 5 cm, %
NS4	0·1% Cu, 1·7—2·8% Mg, 0·6% Si, 0·5% Fe, 0·5% Mn, 0·2% Zn, 0·30% Cr, 0·15% Ti, remainder Al	0 Half-hard	188 230	18 5
B. Plate to BS 1477				
NP8	0·1% Cu, 4·0—4·9% Mg, 0·4% Si, 0·4% Fe, 0·2% Zn, 0·5—1·0% Mn, 0·25% Cr, 0·25% Ti, remainder Al	0	280	12—16
HP30	0·1% Cu, 0·4—1·4% Mg, 0·6—1·3% Si, 0·5% Fe, 0·4—1·0% Mn, 0·1% Zn, 0·3% Cr, 0·2% Ti, remainder Al	W†	200	12—15
C. Extruded sections to BS 1471				
E1C	Min. 99% Al, 0·1% Cu, 0·5% Si, 0·7% Fe, 0·1% Mn, 0·1% Mn	As manufactured	170	18
HE9	0·1% Cu, 0·4—0·9% Mg, 0·3—0·7% Si, 0·5% Fe, 0·3% Mn, 0·1% Zn, 0·1% Cr, 0·2% Ti	0 As manufactured	108 138	13—15 16—18
HE30	0·1% Cu, 0·4—1·4% Mg, 0·6—1·3% Si, 0·5% Fe, 0·1% Zn, 0·4—1·4% Mn, 0·3% Cr, 0·2% Ti, remainder Al	0 As manufactured	107 186	13—15 18
D. Drawn tube to BS 1471				
HT30	0·1% Cu, 0·4—1·4% Mg, 0·6—1·3% Si, 0·5% Fe, 0·1% Zn, 0·4—1·0% Mn, 0·3% Cr, 0·2% Ti, remainder Al	As manufactured	218	12
E. Extruded round tube and hollow sections to BS 1474				
HV9	0·1% Cu, 0·4—0·9% Mg, 0·5% Fe, 0·3—0·7% Si, 0·3% Mn, 0·1% Zn, 0·1% Cr, 0·2% Ti	As manufactured	108	15
F. Castings to BS 1490				
LM4	2—4% Cu, 0·15% Mg, 0·8% Fe, 4—6% Si, 0·3—0·7% Mn, 0·3% Ni, 0·5% Zn, 0·2% Ti, 0·1% Pb, 0·05% Sn, remainder Al	As manufactured	154	2
LM5	0·1% Cu, 3—6% Mg, 0·3% Si, 0·6% Fe, 0·3—0·7% Mn, 0·1% Ni, 0·1% Zn, 0·2% Ti, 0·05% Pb, 0·05% Sn, remainder Al	As manufactured	139 to 154	2
LM6	0·1% Cu, 0·1% Mg, 10—13% Si, 0·6% Fe, 0·5% Mn, 0·1% Ni, 0·1% Zn, 0·2% Ti, 0·1% Pb, 0·05% Sn, remainder Al	As manufactured	124 to 162	2 to 3

* Material in the fully annealed condition
† Material in the solution-treated and naturally aged condition

H

CHEMICAL PROPERTIES

Aluminium has, in general, a far better resistance to normal corrosive agents than iron and steel, owing to the formation of a well-adhering coating of aluminium oxide on the surface, which is non-porous and protects the surface beneath.

The following table gives the rate of attack per year on 99·5% pure aluminium in the presence of various chemical agents at 15°C:

Agent	Rate of attack per year, g/m^2
Still water	Negligible
Running water	Negligible
Steam and oxygen	Negligible
Sea water	Negligible
Dilute acetic acid	72
5% NH_4Cl	145
1% $MgCl_2$	24
5% $MgSO_4$	12
NaOH (any concentration)	Very high
HCl (any concentration)	High

Aluminium is completely resistant to most organic materials, including many organic acids, whose reaction with aluminium is slight even under conditions of high concentration. Hydrochloric acid, and similar non-oxidizing acids, dissolve aluminium readily, but sulphuric acid has little effect when cold and at a concentration below 40%. Cold concentrated nitric acid has little effect, but dilute nitric acid, especially when hot, attacks aluminium readily. Aluminium readily resists atmospheric pollution.

Aluminium is attacked by alkalis, particularly sodium and potassium hydroxides. The rate of attack by calcium hydroxide, calcium oxide and materials containing such substances (wet cement, etc.) is rapid, with hydrogen liberated during the reaction, but the protective film formed during the reactions slows down further attack. This reaction is made use of in the manufacture of gas concrete (Chapter 2). When aluminium is painted, care must be taken that these paints do not contain oxides of heavy metals such as lead, as these tend to cause aluminium to corrode. Most salts, on the other hand, provided they are not alkaline, as for example sodium carbonate, have little effect on aluminium.

Sea water has little effect upon pure aluminium or upon aluminium which has been treated by anodizing. Hard waters, which contain calcium and sodium salts, are somewhat corrosive, and the presence of chlorine increases this. However, attack on aluminium is inhibited. When sea water or hard waters attack aluminium that has not been anodized, the corrosion is nevertheless restricted to the surface, as the body of the metal is protected by the thin film of corrosion products. Aluminium alloys can be classified into six main binary groups, according to the main metals added to aluminium metal.

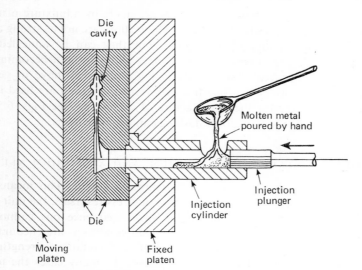

Fig. 6.5   Principle of the pressure die-casting operation
(By courtesy of the Aluminium Federation)

From these binary alloys numerous other alloys are then obtained by adding additional metals in small quantities.

The primary aluminium alloys used in *building* are:

1 *Manganese alloys* contain about 1·2% manganese, are stronger than aluminium and more corrosion-resistant. On the other hand, this alloy is as easy to work and weld as pure aluminium. It is widely used for building purposes, as a cladding and roofing material.

2 *Magnesium* alloys with aluminium are strong, ductile and extremely corrosion-resistant. Such alloys are used for drawing and casting purposes. Magnesium aluminium alloys are particularly useful for tubular constructions, such as scaffolding and the like, owing to their resistance to normal corrosion, and their relatively good fatigue-resisting properties.

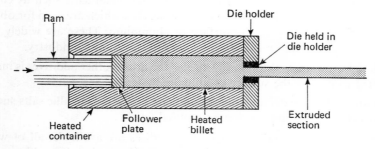

Fig. 6.6   Principle of operation of the extrusion press
(By courtesy of the Aluminium Federation)

3 *Magnesium silicide.* Aluminium alloys are made by admixture of up to
1·5% of magnesium silicide (Mg_2Si). These alloys are not as strong as the
various copper alloys of aluminium, but they can be annealed and
worked much more freely than the copper alloys without becoming too
hard. Their electrical conductivity is high, and they have in general
excellent anti-corrosion properties. Alloys of this group are used in the
building industry for the manufacture of extrusions (see Figs. 6.5 and
6.6), such as window sections, flashings and similar items.

FATIGUE RESISTANCE OF ALUMINIUM AND ALUMINIUM ALLOY

Aluminium and its alloys are subject to vibration fatigue. This means that
when a section is subject to a given number of stress reversals, the unit has a
markedly lower strength than originally. In consequence one cannot use
aluminium and its alloys where fluctuating load conditions are encountered,
except if one designs the unit taking account of the final fatigue strength after
the appropriate number of reversals of stresses. In many cases the tensile
strength of aluminium castings is reduced to between 25 and 35% of its
original value after 10^8 reverses of stresses. This is accentuated at casting
faults where outright failure is then likely to take place (Fig. 6.3).

SURFACE TREATMENT OF ALUMINIUM AND ITS ALLOYS

For many applications, such as roofs, aluminium needs no special surface
treatment. To improve its appearance and also to reduce its reflective properties aluminium can be surface-treated. This is usually carried out by dipping
the object into a chemical solution. The original bright appearance of the
aluminium can also be retained either as such or in the dyed form by using a
process called 'anodizing'.

Chemical processes
The chemical solutions most widely used include:
1 Sodium carbonate, sodium dichromate and metal salts such as copper
   sulphate, nickel sulphate, cobalt chloride, etc., which are used for obtaining coloured films on the surface of aluminium. These are widely used
   for decorative and protective purposes in the building industry.
2 Sodium chromate solution together with various other additives is mainly
   used to protect the surface of aluminium against sea water.
3 Sodium fluosilicate, ammonium sulphate and some metallic salts such as
   nickel sulphate give a coloured mottled finish.

A number of modifications of these processes are available, all of which
produce coatings that are moderately resistant to corrosion, but seldom equal
the electrolytic methods in this respect.

Anodizing

By placing the aluminium object at the anode of an electrolytic cell, and passing a current, oxygen liberated at the anode causes the formation of a relatively thick film of aluminium oxide there, which is smooth and coherent. The electrolyte is usually chromic acid, sulphuric acid, oxalic acid or mixtures of some of these. The films produced are normally either Al_2O_3 or $Al_2O_3 . H_2O$, and the thickness is between 0·003 and 0·035 mm. For exposure to industrial atmospheres the minimum film thickness recommended is 0·025 mm. The films can be coloured by the addition of metallic salts such as nickel sulphate, cobalt chloride, etc., to the electrolyte. The oxide coatings have a very considerable surface hardness which can at times surpass that of either steel or chromium. When the oxide film is obtained by the use of sulphuric acid electrolysis, it is flexible enough for stamping operations.

When it is desired that aluminium sheeting should have as high a reflectivity as possible, a process called the Brytal technique can be used, in which the metal is first treated by immersion in 15% anhydrous sodium carbonate and 5% sodium triphosphate, followed by electrolysis, with the aluminium remaining the anode. The reflectivity of aluminium when treated in this way is 84% and remains at this figure even after exposure lasting for several months. Methods like this are used for treating aluminium foil used for thermal insulation purposes.

Electroplating

Aluminium is sometimes plated by nickel and chromium. This process is carried out in five stages:

1 Degreasing.
2 Electrolytic cleaning.
3 Formation of undercoat.
4 Electroplating.
5 Polishing.

Degreasing is carried out with trichlorethylene vapour, and electrolytic cleaning by using the object as the cathode in a bath of nitric and hydrofluoric acids. The undercoat is usually of zinc, copper or nickel, or a combination of these, which is finished by polishing and attachment of a final layer of chromium by electroplating, usually only 0·025 mm thick.

PAINTING OF ALUMINIUM

In addition to anodizing, the surface of aluminium and its alloys can be made adherent to paint by either mechanical abrasion or by the use of an etch primer, which is a paint with a zinc chromate base. The latter technique is the one most frequently used in the building industry. Paint and enamels used with aluminium must be elastic to compensate for the high coefficient of thermal expansion of the metal.

6.2 Steels and alloy steels

Although nowadays buildings no longer depend on heavy girder and stanchion construction as a framework, as for example the American skyscrapers built in the inter-war years, modern building construction has still many highly sophisticated uses for steel.

In concrete construction, steel is needed for reinforcement and for steel strands and cables in prestressed sections. Steel sheeting, either galvanized or plastic-coated, is used as a cladding panel in modern industrialized buildings. Steel framework components are used for large-span buildings, steel sheet is used as a permanent shuttering, and there are also all types of steel fasteners, galvanized steel ducts, water tanks, grilles and other components, pipings, manhole covers, etc.

Steel has numerous advantages over other metals. It is tough, both in tension and shear, has good elasticity, high creep resistance and reasonable immunity against metal fatigue. Furthermore, there are so many different alloys of steel that it is easy to find one which is tailor-made for the job in hand. Steel is relatively cheap, easy to work and has a high strength to weight ratio. Against all these advantages, steel has only one disadvantage. This is its liability to corrode, unless suitable precautions are taken. This matter will be taken up in Chapter 8.

MANUFACTURE OF STEEL

The first stage of manufacture is the production of iron. The most common ores used are magnetite (Fe_3O_4) and haematite (Fe_2O_3), both of which are usually mixed with impurities of SiO_2, CaO and Al_2O_3. There are also traces of sulphur and phosphorus, both of which are harmful to the final product. Smelting is carried out in a blast furnace, the feed consisting of ore, metallurgical coke and limestone. Air at 700°C, enriched with 28% oxygen, is passed through and the reactions taking place are as follows:

$$2C + O_2 \rightarrow 2CO$$
$$Fe_3O_4 + 4CO \rightarrow 3Fe + 4CO_2$$
$$Fe_2O_3 + 3CO \rightarrow 2Fe + 3CO_2$$
$$CO_2 + C \rightarrow 2CO$$
$$\underset{\text{(impurities)}}{CaCO_3} + \underset{}{SiO_2} \rightarrow \underset{\text{(slag)}}{CaSiO_3} + CO_2$$

The molten iron runs to the bottom, with a layer of molten slag, which contains other impurities apart from SiO_2, on top. The iron, when drawn off, is called pig iron.

Steel is made from the pig iron by a variety of processes, called by such names as: Bessemer, open-hearth, electric-arc, Kaldo, LD, Rotary, Ajax, etc. In all these processes the impurities contained in the pig iron are eliminated, and instead useful additions are made to the melt to produce a metal suitable for the job for which it is going to be used.

There are two ways of getting rid of impurities. One is to pass air or pure oxygen through the melt to oxidize waste constituents, e.g.

$$2C + O_2 \rightarrow 2CO$$
$$C + O_2 \rightarrow CO_2$$
$$S + O_2 \rightarrow SO_2$$
$$Si + O_2 \rightarrow SiO_2$$

The other is to add quantities of iron oxide to the melt, which reacts as follows with impurities:

$$5FeO + 2P \rightarrow P_2O_5 + 5Fe$$
$$FeO + C \rightarrow CO + Fe$$

The oxides are either blown off, or are removed in the form of a liquid slag from the top of the molten metal. Pig iron, which has had all its impurities removed, becomes what is termed 'wrought iron', which is useful for such purposes as the manufacture of crane hooks, and for the production of ornamental architectural features such as wrought iron gates, fencing, etc. Its tensile strength is only 340—360 MN/m² with a Brinell hardness of 75—80. Only about 2% or less of the ferrous metals produced each year are, in fact, wrought iron.

PROPERTIES OF PURE IRON

Pure iron has a density of 7·86 kg/dm³ at 20°C and a melting-point of 1,535°C, with a boiling-point of 3,235°C. The thermal conductivity is 48·5 W/m.degC. Its coefficient of linear expansion is equal to $1·45 \times 10^{-5}$ degC^{-1} and its electrical resistivity at 20°C is equal to 0·041 Ω.cm. The specific heat of iron is equal to 0·458 kJ/kg.degC. Chemically it is readily attacked by most acids but not by alkalis, which tend to form a passive film on the surface of the metal. Iron also has some resistance to concentrated nitric acid which makes the metal passive. Iron reacts with steam at a temperature of 570°C and above to produce iron oxide and hydrogen. At temperatures below this, reactions also proceed but are much slower. Iron is also attacked by ammonia at high temperatures.

These reactions, and also the more common corrosion reactions, which will be dealt with fully in Chapter 8, depend very markedly upon the additives made to iron.

PRODUCTION OF STEEL

The main chemical elements which are added to the purified pig iron are the following: carbon, manganese, chromium, nickel, titanium, molybdenum, vanadium and tungsten.

This list is by no means comprehensive, however, and numerous other metals are added to achieve certain results. Aluminium is also added to the melt, but the purpose is to remove small quantities of iron oxide by inducing the following reaction:

$$3FeO + 2Al \rightarrow 3Fe + Al_2O_3$$

the aluminium oxide being removed with the slag.

BINARY CARBON/IRON ALLOYS

The basic steel alloy is a combination of iron and carbon, shown as a phase diagram in Fig. 6.7. The important region is the austenite region which is a solid solution of cementite (Fe_3C) in iron. If steel is cooled slowly, the austenitic state changes at room temperature to the pearlitic/ferritic stage which is quite soft. When steel is in this form it is easily worked and shaped. If red-hot steel is, however, quenched by being suddenly immersed into cold water or cold oil, the austenite changes into a glass-hard state called martensite. This material is in a metastable state, and unless stabilized by the addition of such metals as manganese, titanium, chromium or vanadium, tends to change to the

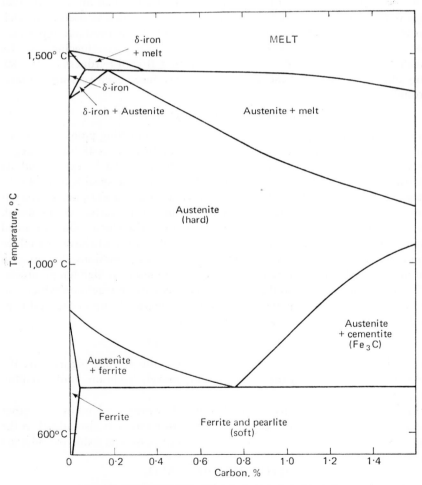

Fig. 6.7   The phase rule diagram for iron–carbon mixes

ferrite/pearlite state with consequent loss of hardness and strength. But pure martensite is rather brittle and for this reason objects are tempered after quenching. Tempering consists of heating the object gently to convert a small portion of the martensite into ferrite/pearlite until the object has lost its basic brittleness, followed by air cooling.

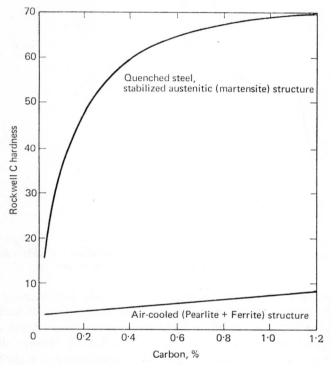

Fig. 6.8   Relationship between the hardness of steel and its carbon content

CASE HARDENING

Very hard objects are usually brittle, while tough objects with a high Izod impact figure are soft. For many tools, the shank is constructed from a medium carbon steel, with the tip consisting of pure martensite high carbon steel. This is done by the process of shaping the tool first and then immersing the red-hot tip into a carburizer, usually carbon, but frequently also organic compounds containing carbon and nitrogen. Quenching then produces a glass-hard working surface.

MILD STEEL

Steels with a very low carbon content (below 0·20%) are softish and ductile (see Fig. 6.8). The mild steels cover the entire range from virtually wrought

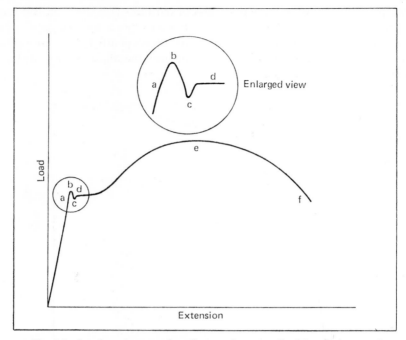

Fig. 6.9   Load versus extension diagram for normalized low carbon steel
(By courtesy of the British Steel Corporation)

iron up to and including some structural steels. Mild steels account for some
90% of the output of the steel industry. The main forms of mild steel of
interest to the building industry are steel sheeting, steel pipes, rivets, reinforcement bars and mesh for concrete, etc. The relationship between the carbon
content of steel and the tensile strength can be seen from Fig. 6.10. Mild steel
cannot be hardened by quenching and tempering.

STRUCTURAL STEEL

This is used for such purposes as beams, girders, channels, etc., and is in
general made with a carbon content of between 0·15—0·25%. It also includes
a number of other constituents, up to 1·50% manganese, up to 0·50%
chromium and traces of other elements. Sulphur and phosphorus must always
be less than 0·05% each. The tensile strength ranges specified by BS 968:1962
for high yield stress, welding-quality structural steels are between 50·4 and
61·4 kgf/mm² (493—602 MN/m²) with a yield stress of 34·6—36·2 kgf/mm²
(340—356 MN/m²). The elongation on a 200-mm test piece varies between
15 and 18%. The impact test according to BS 131:2 on samples 50·8 mm
thick shall produce averages above 2·76 kg.m at −15°C, with no value below
2·07 kg.m.

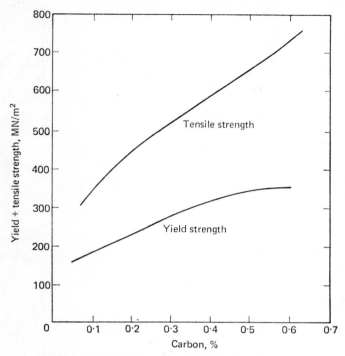

Fig. 6.10   Yield and tensile strength of steel containing
0·1 % Si, 0·4 % Mn, 0·04 % S and 0·17 % P

The flexibility of structural steel is tested by bending over a steel plate. Plates up to 25 mm thick should be capable of being folded over with a diameter of twice the thickness, and plates in excess of this thickness to a diameter three times the thickness.

HIGH CARBON TOOL STEELS

These are steels which contain between 0·5 and 2·0 % carbon, together with traces of other elements. They are somewhat cheaper than the alloy steels, with which they compete directly, but are not quite as servicable as the latter. In the building industry high carbon tool steels are used for hammers, drills, cold chisels and other purposes where intermediate properties between the alloy steels and the structural and mild steels are required.

ALLOY STEELS

These are steels with varying carbon content to which elements such as chromium, manganese, nickel, tungsten, vanadium, silicon, molybdenum, niobium,

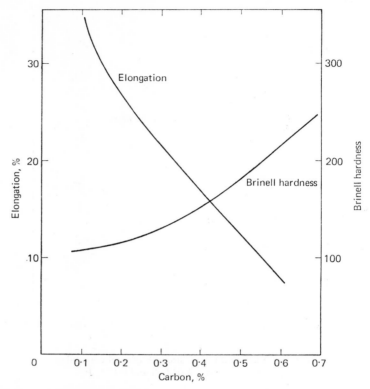

Fig. 6.11   Brinell hardness and percentage elongation of steel containing
0·1 % Si, 0·4 % Mn, 0·04 % S and 0·2 % P

lead, copper and many others are added in larger or smaller quantities to
obtain alloys which have one or more of the following characteristics:

1 Good strength.
2 Good elasticity.
3 Good quench hardness.
4 Low rate of hardening during cold working.
5 Good abrasion resistance.
6 Resistance to warping or cracking.
7 Good high and low temperature properties.
8 Resistance to rusting.
9 Resistance to named chemicals.

Manganese
This element is present in nearly all steels and has the function of combining
preferentially with oxygen and sulphur, as well as increasing the tensile strength
and hardness (see Fig. 6.12). The ductility is, however, reduced somewhat.

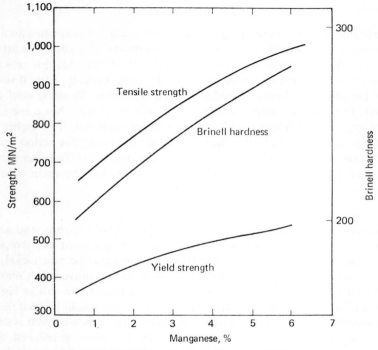

Fig. 6.12   Effect of percentage manganese content on 0·55% C steels

Chromium and nickel

These are the most common alloying elements and are used in small quantities for producing stronger and tougher forms of steel, by stabilizing the austenitic state, and in larger quantities for making the whole range of stainless steels mentioned below. Chromium and nickel alloy steels are used in the building industry for high tensile steel bolts, for the manufacture of steel wire and cable used for prestressing and post-tensioning purposes, and for the construction of springs.

Tungsten, vanadium, molybdenum and cobalt

These elements are used for modern cutting tools to prevent the steel from softening as high operating temperatures are attained. Modern cutting tools can be heated almost to red heat without losing their hardness. The following are typical formulations for various types of high-speed alloy steels:

High tungsten steel	0·65% C, 19% W, 4% Cr, 1% V
Low tungsten steel	0·75% C, 14% W, 4% Cr, 2% V
Molybdenum steel	0·85% C, 1·5% W, 4·25% Cr, 1·0% V, 8% Mo
Cobalt steel	0·75% C, 18% W, 4% Cr, 2% V, 8% Co

Rest Fe

The fundamental basis underlying stainless steel is the fact that when an alloy containing a high percentage of chromium is exposed to a corrosive atmosphere, an adherent film of chromium oxide (Cr_2O_3) is formed; this acts as a passive surface, preventing the corrosion of the metal beneath. Even if such a film is broken, it re-forms, provided oxygen is present. Stainless steel can, however, fail if the chromium oxide film should be broken, but there is no oxygen around which can enable the film to be reformed. The higher the percentage of chromium in the alloy, the better the protective action on the steel. Nickel serves to improve the stability of the oxide film. There are almost 100 different types of stainless steel on the market, but the main types of interest to the building industry are as follows.

Martensitic stainless steels

The martensitic stainless steels contain about 0·1—0·4% carbon and about 14% of chromium. They are very hard and can be quenched and tempered like normal tool steels. Other elements, such as manganese and nickel, are added to stabilize their structure, but this is by no means universal. A martensitic stainless steel with 0·35% carbon can have a tensile strength as high as 850 MN/m² and a Brinell hardness up to 230. This type would be used for the manufacture of stainless steel cutting tools. When the carbon content is somewhat lower the material is not quite as hard, but brittleness is reduced. Such forms of martensitic stainless steel are employed for cables and wires, jointing brackets, etc. Excessive cold-working reduces the corrosion resistance of martensitic stainless steels.

Austenitic stainless steels

The austenitic stainless steels are by far the most popular, and are based upon an approximate formulation of 18% chromium and 8% nickel. Hence the term 18/8 for such a type of stainless steel. The austenitic structure, which is the solid solution of iron carbide and all the metal additives in iron, is stabilized at all temperatures, producing a soft and flexible product. During cold-working some of the austenite is converted to martensite and this does harden the steel and increases its strength.

The normal grade of austenitic stainless steel used for such purposes as fume hoods, decorative cladding, architectural features, etc., has a carbon content of up to 1·0% a yield strength of 270 MN/m² and a tensile strength of around 750 MN/m². The elongation varies between 30 and 73% depending upon carbon content and the Brinell hardness between 160 and 200. The impact strength is around 15—18 kg.m (Izod impact).

Stainless steel with a better corrosion resistance than normal, to be used under circumstances where acid fumes abound, has a very low carbon content (0·05%), together with 0·6% Ti, 3·5% Mo and 2·0% Cu, in addition to the usual 18% Cr and 8% Ni.

Ferritic stainless steels

These are very soft stainless steels with a low tensile strength but a high elongation figure. They contain 0·1 % C and between 15 and 20% chromium. The tensile strength is about 400—500 MN/m², elongation on a 5 cm long sample is up to 40%, but their resistance to impact is very poor, with Izod impact figures as low as 1—2 kg.m being quoted. This stainless steel is used almost exclusively for one purpose, the production of pressed and deep-drawn articles such as kitchen sinks, bowls and similar products.

HEAT RESISTANT ALLOYS

A large number of high temperature alloy steels is on the market, able to withstand temperatures of up to 800°C and above. A typical formulation of such a steel is 0·3% C, 19% Cr, 9% Ni,ʳ1% Mn, 0·6% Si, 1·2% Mo, 2% Ti and 0·3% Al. This alloy can resist a rupture stress of 130 MN/m² for 1,000 hr at a temperature of 730°C, and a rupture stress of 70 MN/m² for 1,000 hr at a temperature of 815°C.

Literature Sources and Suggested Further Reading

1. AITCHISON, L., and PUMPHREY, W. I., *Engineering Steels*, Macdonald and Evans, London (1953)

2. ALUMINIUM FEDERATION, LONDON, Technical Brochures (1963–1968)

3. 'Aluminium in Building', Symposium, London, 9 and 10 July 1959, *Proc. Royal Inst. of Brit. Architects*

4. BASHFORTH, G. R., *The Manufacture of Iron and Steel*, 4 volumes, Chapman and Hall, London (1951–1964)

5. BENBOW, W. E., *Steels in Modern Industry*, Iliffe, London (1951)

6. CHATER, W. J. B., and HARRISON, J. L., *Recent Advances with Oxygen in Iron and Steel Making*, Butterworth, London (1964)

7. DENNIS, W. H., *Metallurgy of the Ferrous Metals*, Pitman, London (1963)

8. GALE, W. K. V., *The British Iron and Steel Industry*, David and Charles, Newton Abbot (1967)

9. GODARD, H. P., et al., *The Corrosion of Light Alloys*, Wiley, New York (1967)

10. HORN, K. R. VAN, *Aluminium*, 3 volumes, American Society for Metals, Metals Park, Ohio (1967)

11. KIRKALDY, J. S., and WARD, R. G., *Aspects of Modern Ferrous Metallurgy*, Blackie, Edinburgh (1964)

12. LANCKER, M. VAN, *Metallurgy of Aluminium Alloys*, Chapman and Hall, London (1967)

13. OSBORNE, A. K., *Encyclopaedia of the Iron and Steel Industry*, Technical Press, London (1967)

Chapter Seven Copper, Lead and Zinc

Apart from iron and steel on the one hand, and aluminium on the other, copper, lead and zinc are the metals most widely used in construction.

7.1 Copper

OCCURRENCE AND SMELTING PROCESSES

Most of the common copper ores of commerce contain a good deal less than 4% of copper metal. The copper is usually found as a sulphide in intimate admixture with iron sulphides and many other compounds, the bulk of the ore consisting however of waste rocks. The total annual production of copper

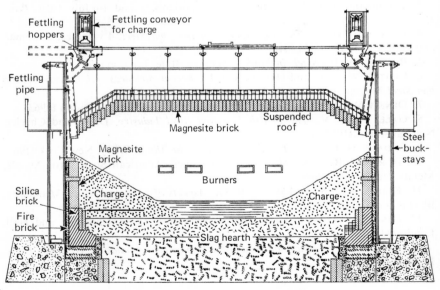

Fig. 7.1   Diagrammatic cross-section of a reverberatory smelting furnace
(By courtesy of the Copper Development Association)

is around 5 million tons, and about a third of this comes from the USA. Other countries, which each have a share of around 15% of the total are USSR, Zambia and Chile, the rest being produced by Canada, the Congo, Japan, Australia and various European countries.

The ores are first ground and then treated by froth floatation. The copper compounds are not wetted by the liquid employed and are carried over, while the siliceous materials sink to the bottom of the cell and are disposed of. The froth which is collected is 'copper concentrate' and is next heated in a reverberatory furnace (see Fig. 7.1). On heating, two melts are formed, a high-density mixture of copper 'matte' and a low-density mixture of molten slag. The matte is separated from the slag and transferred to a converter, which acts somewhat like the well-known Bessemer converter used in the steel industry. It is a large steel drum lined with refractories and has air inlets to permit the oxidation of the copper sulphides and other compounds:

$$2CuFeS_2 + 2SiO_2 + 4O_2 \rightarrow Cu_2S + 2FeSiO_3 + 3SO_2$$
$$Cu_2S + 2O_2 \rightarrow 2CuO + SO_2$$
$$Cu_2S + 2CuO \rightarrow 4Cu + SO_2$$

The slag, which contains the iron silicates, is removed and the crude copper, which is about 99% pure and is known as 'blister copper' because of its appearance, is cast into slabs for further treatment.

Copper can either be 'fire-refined', which produces a grade suitable for the manufacture of alloys, or 'electrolytically refined', which produces the high-purity grades needed for many electrical and other purposes.

PROPERTIES OF COPPER

The three main grades of copper used in engineering are:

1 *High conductivity copper*, which has a purity of 99·9% and above. It is used for electrical purposes and also where high thermal conductivity is of importance. Pure copper has an electrical resistance at 20°C of $1·7241 \times 10^{-6}$ Ω.cm which is affected considerably by even minute additions of other materials and temperature changes. The thermal conductivity of the purest forms of commercial copper is equal to 393 W/m.degC. The density of high conductivity copper is equal to 8·89 kg/dm³ and its melting-point is 1,083°C. In its purest form copper forms a single phase only, except for small quantities of cuprous oxide (Cu_2O) present. It is these traces of cuprous oxide which tend to form grain boundaries in the copper structure.

2 *Refined tough-pitch copper*. There are several grades of this type of refined copper, which contains small quantities of various impurities. This type of copper is the one most widely used.

3 *Arsenical copper*. Arsenical copper contains up to 0·5% arsenic. Although

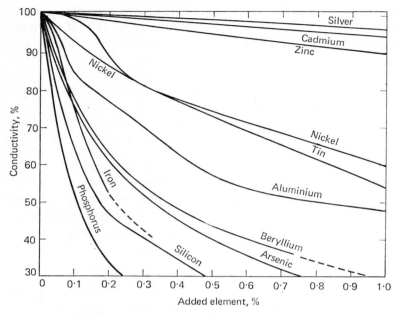

Fig. 7.2 Effect of added elements on the electrical conductivity of copper

the electrical and thermal conductivities of the material are still high in comparison with other metals, they are much lower than those of high-conductivity grades of copper. Arsenical coppers have a considerable increase in work-hardened strength and hardness over other forms of copper, and in addition, the endurance limit, as determined by fatigue testing, is raised. Arsenic in copper also raises the annealing temperature by about 100°C and improves the strength of the copper at higher temperatures.

MECHANICAL PROPERTIES OF COPPER

The figures given in Table 7.1 apply to normal refined, tough-pitch copper, which is the most usual form encountered in the building industry.

The tensile and other strengths of copper are often affected by the previous history of working of the object in question, and heavily cold-worked copper

TABLE 7.1

Condition	Tensile strength, MN/m²	Elongation on 5 cm before break occurs, %	Diamond pyramid hardness, HV
As cast	155—165	25—30	40— 45
Cold-worked	310—385	5—20	80—115
Annealed after cold-working	216—247	50—60	45— 55

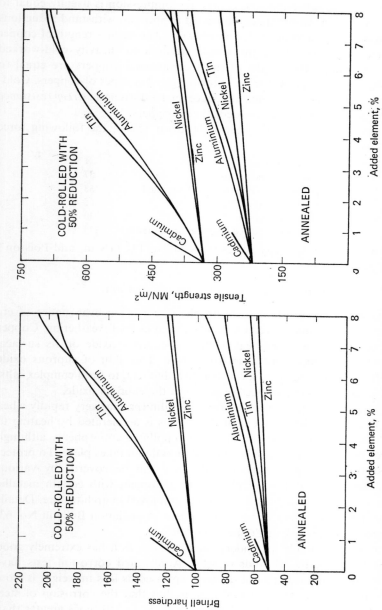

Fig. 7.3 Addition of other elements to harden and strengthen copper. Brinell hardness and tensile strength of copper alloys (By courtesy of the Copper Development Association)

can have a tensile strength as high as 400 MN/m². The linear expansion of copper equals $2 \cdot 1 \times 10^{-6}$ degC⁻¹.

The limit of proportionality of copper in compression is usually equal to the value of tension. Good quality annealed copper can withstand reductions of thickness of 90% and more during working. The shear strength of copper is approximately one-half of the tensile strength for heavily cold-worked material. Izod impact strengths of most commercial coppers are equal to 5·5—6·9 kg. m, the upper value being for oxygen-free types of coppers. Cold-working, presence of oxygen, etc., all reduce the Izod strength. *Creep resistance* is highest for drawn phosphorus, deoxidized coppers.

To achieve a creep of 0·01% per 1,000 hr at 150°C, the following force (MN/m²) is needed:

Electrolytic annealed	21
Phosphorus deoxidized annealed	37
Phosphorus deoxidized arsenical annealed	58
Electrolytic drawn (84%)	65
Phosphorus deoxidized drawn (84%)	230
Phosphorus deoxidized arsenical drawn (84%)	275

Young's modulus of elasticity varies from 110 to 132 GN/m² and Poisson's ratio is 0·35.

CHEMICAL PROPERTIES OF COPPER

Copper is extremely resistant to atmospheric corrosion even under severe conditions and is virtually everlasting when used as roof weathering. Copper forms a continuous and self-healing film of cuprous oxide on its surface, which, unlike rust, does not absorb moisture. This skin of cuprous oxide reacts with water and acidic gases such as SO_2 and CO_2 to form complex salts, which also have the same protective action as the cuprous oxide.

Polished copper acquires such a protective film (patina) very rapidly when exposed to air. The development of these films is accelerated by heating to 150°C. Polished copper is rapidly tarnished in polluted atmospheres, although afterwards the action stops and no corrosion in depth takes place. To protect copper surfaces against surface tarnish, they can be covered by various lacquers. Surfaces can also be coloured by treatment with certain metallic sulphides, a practice widely used for copper employed in architecture. Details can be obtained from the Copper Development Association Bulletin No. 63, 'Architectural Metalwork in Copper and Alloys'.

Copper is very suitable for underground services as it has extremely good resistance to attack by all kinds of soils. Accelerated corrosion tests have shown that the rate of corrosion of steel expressed as loss in weight is from 13 to 30 times greater than that of copper, but that the corrosion of steel when expressed as maximum pit depth is from 74 to 250 times greater than that of copper. The worst cases of corrosion of copper objects buried in the soil occur with soils that are badly drained.

Corrosion troubles, however, occur where copper and steel pipes are joined together, because copper induces corrosion in steel objects connected to it due to electrochemical action (see Chapter 8). Such damage to steel pipes can be prevented by ensuring that the steel in contact with the copper is either completely covered, or completely bare. In the former case, water cannot get at the iron surface so that corrosion cannot start, while in the latter case the e.m.f. effect is diluted and corrosion is only slight. Really heavy corrosion takes place when the steel is covered, but small pin-holes are left. In such circumstances the steel pipe is perforated in a very short time. When steel and copper pipes are joined it is recommended that a band of steel pipe 30 cm wide from the join should be given good surface protection, and from then onwards the steel pipe should be left uncovered.

CHEMICAL RESISTANCES

Copper has good resistance to the following:

1 *Acids*. Acetic, carbonic, citric, formic, oxalic, dilute sulphuric and sulphurous acids. There is in many cases some corrosion effect when air is present.

2 *Alkalis*. Dilute and concentrated sodium and potassium hydroxide solutions.

3 *Salts, etc.* Most salts with the exception of those of iron and mercury, domestic water supplies, sea water and industrial liquors.

The following solutions attack copper: concentrated sulphuric acid, nitric acid at any strength, ammonia, acid chromate solutions, ferric salts, mercury salts, perchlorates and persulphates.

COPPER ALLOYS

When small quantities of arsenic, silver, cadmium, chromium, tellurium, beryllium or silicon, etc., are added to copper, considerable modifications in properties are obtained. For example, 1% cadmium adds to the strength of copper wire, while a small quantity of beryllium forms an alloy, which when suitably heat-treated, is actually stronger than steel.

Brasses

These are alloys of copper and zinc. Various proportions of the two metals are used for different purposes but one of the most popular is the 60/40 'leaded' brass, used for the manufacture of a wide range of fittings in the building industry.

Bronzes

These are alloys of copper and tin, usually with some other elements such as phosphorus added. Normal phosphor bronze, which contains 5—6$\%$ tin and traces of phosphorus, is used for the manufacture of woven-wire gauzes for filters and screens. Bronze containing about 12% of tin is used for such architectural purposes as ornamental doors and other bronze fittings.

Copper – nickel – silver
These alloys are often used for various architectural features such as shop windows, counters, showcases and door fittings. These fittings can be left either untreated or, as happens more frequently, they are coated over by a layer of chromium plating.

USES OF COPPER AND ITS ALLOYS IN THE BUILDING INDUSTRY

Copper is mainly used for the following purposes:

1 *As piping.* The vast majority of copper pipes used for domestic hot and cold water supplies, panel heating and sanitary purposes are seamless ones, produced by an extrusion process. Copper pipes have excellent resistance to corrosion, are easy to bend and to join, and also, when used as water pipes, have good frost resistance. It is estimated that about 1,000,000 tons of tubes made from copper or copper alloys are manufactured annually in the United Kingdom, most of which are used by the building industry.

2 *As roof covering, flashing, etc.* When copper is employed for purposes of protecting a building against rain, it is usually used in the form of rolls or coils of strip, i.e. flat sheeting 50 cm or less in width. Such material is readily fabricated into gutters, rainwater heads, ventilators, sill weathering, etc. Copper sheet is also fabricated into hot water tanks and domestic boilers. For roofing, copper is laid in the form of flat sheets or rolls of strip which can be cut *in situ*. Copper is often also used for covering curved or irregularly shaped surfaces such as domes, a notable example of a recent contract of this type being the copper dome of the London Planetarium.

7.2 Lead

SMELTING AND REFINING OF THE METAL

Lead is found in nature in the form of its sulphide, PbS (galena), its carbonate $PbCO_3$ (cerrusite) and its sulphate $PbSO_4$ (anglesite), the first of these, galena, being the most common. Lead ores are usually pretreated by being roasted and are then fed into a blast furnace with an admixture of coke. After sintering lead is formed by the following reactions:

$$2PbS + 3O_2 \rightarrow 2PbO + 2SO_2$$
$$PbO + C \rightarrow Pb + CO$$

The lead is recovered from the bottom of the furnace, while most of the impurities are collected in the lighter slag.

Purification of lead is carried out either by adding zinc to it, which removes silver (the Parkes process), or by electrolysing lead in a solution of lead fluosilicate and hydrofluosilicic acid. The latter is called the Betts process,

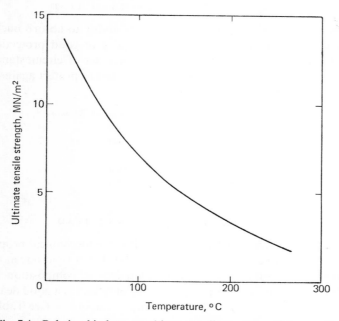

Fig. 7.4   Relationship between ultimate tensile strength and temperature for desilverized lead

which produces pure lead bars at the cathode of the cell. The insoluble sludge which adheres to the crude lead anodes contains nearly all the impurities except tin and some antimony, which can, however, be removed by passing oxygen through the lead melt.

PHYSICAL PROPERTIES OF LEAD

The density of pure lead at 20°C is given as 11·344 kg/dm³. The material is soft and has a hardness on the Moh scale of 1·5, contrasting with a figure of 6 for zinc and 8 for copper. The melting-point of pure lead is 327·4°C and it has a thermal conductivity of 35 W/m.degC at room temperature.

The thermal expansion figures for lead are as follows:

Linear expansion: $2·95 \times 10^{-5}$ degC^{-1}

Cubic expansion: $8·4 \times 10^{-5}$ degC^{-1}

These thermal expansion figures are greater than the equivalent figures for copper, iron and other common metals.

The electrical resistance of lead is rather higher than that of many other metals and amounts to 20.63×10^{-6} Ω.cm at room temperature.

RADIOACTIVITY PROPERTIES OF LEAD

One of the most useful properties of lead is its ability to absorb nuclear and X-ray radiation. Although other materials also give good protection, lead provides the thinnest protective layer for a given set of circumstances. The following table gives the minimum thickness of lead to protect against X-rays produced by different peak voltages.

Peak voltage, kV	Thickness of lead shield, mm
100	1·5
200	4·0
300	9·0
400	15·0
500	22·0

MECHANICAL PROPERTIES OF LEAD

It is difficult to give accurate values for most of the mechanical properties as lead behaves sometimes more like a plastic solid than a rigid one at ordinary temperatures. In addition, very slight differences in composition have an enormous effect upon the general values. Very pure lead is a good deal weaker and softer than commercial lead which contains impurities (see Table 7.2).

TABLE 7.2

	Type of lead*		
	A	B	C
Ultimate tensile strength, MN/m²	13·6	19	11·1
Yield point at 0·5% elongation, MN/m²	5·9	11·3	4·9
Elongation, %	64·2	50·2	68·6
Brinell hardness No. 1, 1 hr after casting	4·2	5·5	2·9

* A—South Eastern Missouri desilverized lead
   B—Chemical lead (rather impure)
   C—Very pure lead with 99·99943% Pb

The tensile strength of lead depends considerably upon the temperature as can be seen from Fig. 7.4. Young's modulus for lead varies a good deal, but a usually accepted figure at 20°C is of the order of 18 GN/m², while the coefficient of compressibility of lead at 20°C is 2.33×10^{-6} per 10^6 bar. Lead is very malleable and flexible. Although it hardens temporarily when cold-worked, it regains its original softness simply by standing at room temperature.

Lead has a rather low fatigue limit, which is raised by the addition of small quantities of tin, antimony or cadmium. After 10 million reversals of stress, 99·99% lead in the rolled state has a fatigue limit of 2·6 MN/m². The initial addition of just 0·3% cadmium raises this to 7·3 MN/m².

CHEMICAL PROPERTIES OF LEAD

Chemical resistances are as follows:

1 *Sulphuric and hydrochloric acids*. Good resistance at almost all concentrations when cold, less when hot. The resistance also depends upon the nature and quantities of impurities present.

2 *Nitric acid*. Lead is attacked rapidly by even very dilute nitric acid. Organic acids readily attack lead even when dilute.

3 *Alkalis*. Good resistance to dilute solutions. Stronger alkalis attack lead readily because they form plumbates.

Oxygen in the presence of moisture oxidizes the surface of lead readily. This process takes place particularly readily when there are no dissolved salts present which are able to cover the surface of the lead with a protective film. Lead is attacked by fluorine and chlorine. Nearly all lead salts, except the nitrate and acetate, are virtually insoluble in water.

ATMOSPHERIC CORROSION

When lead is freely exposed to the atmosphere, it is extremely resistant to corrosion due to the readiness with which the surface becomes covered by an impervious film of oxides, carbonates and sulphates. Tests carried out on various metals to compare their resistance to an industrial atmosphere showed that the relative weight losses after seven years' exposure were as follows:

Mild steel	2,609
Wrought iron	1,703
Zinc	137
Nickel	120
Nergandin brass	100
Copper	47
Aluminium	29
Tin	18
99·96% lead	22
Antimonial lead (1·6% Sb)	3

Lead tends to corrode when in contact with rotting timber due to the fact that organic acids are liberated which then readily attack the metal.

SOIL AND WATER CORROSION

Buried lead objects corrode at a rate which is markedly dependent upon the nature of the soil with which they are in contact. Humus and cinders have a greater corrosion effect than either sand or clay. Corrosion may occur due to the effect of soft water, from water carrying calcium hydroxide and from various acidic substances in the soil.

When flowing through lead pipes, hard water is less likely to attack the lead than soft water and it has been found that carbonates, bicarbonates and silicates in solution all have some protective effect. Nitrates, nitrites and ammonium salts, however, increase the rate of corrosion of lead as does the presence of oxygen and carbon dioxide in the water which is being carried. Pure gas-free de-ionized or distilled water does not attack lead nor has chlorination of water up to 0·002 ppm any appreciable effect upon the plumbo-solvent power of water. Lead has been found to be extremely resistant to attack by sea water, with a rate of attack of about 2 mg/dm^2 per day on soft lead bars. The rate of attack on lead is increased when the metal is in contact with copper owing to galvanic action on corrosion processes. This is, however, counteracted by a heavy corrosion product layer on the surface of the exposed lead.

LEAD ALLOYS

A large number of alloying elements is added in smaller or larger quantities to lead in order to modify its properties. Antimony, arsenic, cadmium, calcium, tellurium and tin all have the effect of increasing the tensile strength of lead and improving its hardness, fatigue resistance, etc. Silver alloys are used on account of their good creep resistance and workability, tellurium and copper improve the corrosion resistance, while the addition of selenium reduces the electrical resistivity of lead.

USES OF LEAD IN THE
BUILDING AND CONSTRUCTIONAL INDUSTRIES

Water pipes and other plumbing components

Lead is used for the construction of service mains for the transportation of domestic water and competes in this field with galvanized steel, and copper and its alloys. Lead is also widely used for traps and pipes in soil and waste systems, ventilating pipes and gas pipes. As a piping material lead is very durable and easily worked and installed, but is both expensive and heavy when used for high pressures. When lead pipes are used for the transportation of drinking water great care must always be taken to avoid contamination. In general, a lead content in excess of 0·5 ppm is likely to have considerable and permanent adverse health effects, and it is agreed by most authorities that a maximum lead content below 0·1 ppm should be specified for consumption water. The presence of silicates in the water also has the effect of reducing chemical attack on the lead. For a given flow rate of water, lead pipe has a smaller diameter and a greater wall thickness than equivalent galvanized steel pipes. Lead water pipes are subject to bursting when frozen, although this effect is diminished if tellurium is used in admixture. Ordinary lead pipes may be frozen five times before they burst, while tellurium lead pipes can be frozen twelve times under identical circumstances. In addition, tellurium lead pipes can be made up to 35% lighter than lead pipes designed for the same purpose.

Lead should never be installed in direct contact with concrete, but when for example, concrete has to be lined with lead, an intermediate layer of asphalt should be in position. Lead is widely used as a caulking material for making connections in iron pipe with hub and spigot joints. Often the lead is used in the form of wool, which is hammered into place. Lead can be welded directly but normally piping is soldered and the joints are then wiped. In order to obtain the necessary slow hardening mix to permit wiping, the plumber's solder consists of 60% lead, 38% tin and 2% antimony.

Lead roofing

Lead competes directly with copper as a roofing material. Its advantage is superb durability but it also has a number of disadvantages. Lead roofing is expensive and because of its considerable weight, the supporting structure has to be made stronger. Allowances have to be made for thermal movements. Terne plating, which is iron sheeting that has been dipped into a bath containing molten lead–tin mixtures is used for roofing purposes. The material has some advantages over galvanized roof sheeting from the corrosion resistance point of view. In addition, lead-coated copper is also being used for roofing purposes, the lead being deposited either electrolytically or applied by dipping and rolling.

Flashings and trimmings

Lead flashings are used as weatherproof joints between chimney and wall, along window sills, door frames, for roof ridges and valleys, gutters and leaders. Lead is also extensively used as a damp-proof course. For all these purposes lead tends to be rather more expensive than alternative materials but has the advantage of ease of installation and durability.

Antivibration mats

Lead-asbestos antivibration pads are used to offset vibration in buildings caused by heavy traffic, and in preventing the vibration of moving machinery from being transmitted to the surrounding building. The mats contain a centre layer of steel surrounded by two sheets of asbestos, and encased in lead, 3·2 mm thick. Such pads can sustain unit loads of up to 2 MN/m².

Sound-proofing

Sound insulation provided by a wall interface is expressed by the following equation:

$$\text{Insulation in decibels} = 18 \log_{10} G + 12 \log_{10} f - 25$$

where G (kg/m^2) is the mass density of the material and f (Hz) is the frequency of sound. Because lead has such a high density, it can be used for the sound insulation of light frame walls. A useful sound-resistant type of interface consists of twin sheets of lead with a density of 20 kg/m^2 some distance apart, being supported by a lightweight framework.

7.3 Zinc

SMELTING AND REFINING OF THE METAL

The most important source of metallic zinc is zinc sulphide (ZnS) or blende, which is found in association with lead sulphide, iron pyrites and other constituents such as cadmium and sometimes even gallium, thallium and germanium sulphides. Calamine ($Zn(OH)_2.SiO_2$) and Smithsonite ($ZnCO_3$) are other naturally occurring compounds of zinc. In general, most zinc ores contain less than 5% of zinc, and before smelting operations can be started it is necessary to concentrate these ores to have a zinc content of between 35 and 60%. This is usually carried out by grinding the ore, followed by ore separation using a float and sink method, or various elutriation techniques.

Zinc sulphide cannot be reduced directly to metallic zinc but must first of all be roasted, with the formation of the oxide.

$$2ZnS + 3O_2 \rightarrow 2ZnO + 2SO_2$$
$$(\Delta H = -925 \text{ kJ})$$

During the roasting process the ore is sintered to form hard lumps of oxide from the zinc concentrate which is obtained as a powder from the classifier.

The final stage is the reduction of the zinc oxide and the distillation of the metal obtained. This is carried out by heating the oxide with carbon or carbon monoxide, or both. The reactions which take place are the following:

$$ZnO + C \rightarrow Zn + CO$$
$$(\Delta H = +238 \text{ kJ})$$
$$ZnO + CO \rightarrow Zn + CO_2$$
$$(\Delta H = +65 \cdot 5 \text{ kJ})$$
$$CO_2 + C \rightarrow 2CO$$
$$(\Delta H = +179 \text{ kJ})$$

PROPERTIES OF ZINC

The zinc crystal lattice is a closed packed hexagonal type, and the metal has a density of 7·126 kg/dm^3 at 27°C. The melting-point of pure zinc is 419·5°C and its boiling-point is 905·7°C, which is a lower figure than the *melting-points* of very many metals. Zinc shows no allotropy, which means that there are no phase changes, as are experienced with most other metals as the temperature is raised or lowered. The thermal conductivity of zinc is poor, and the figure quoted for it at atmospheric temperature is equal to $1\cdot12$ W/m.degC, which is about 26% of that of silver and 28% of that of copper. The linear thermal expansion of zinc is equal to $3\cdot95 \times 10^{-5}$ degC^{-1}, although there are some differences depending upon whether the measurements are made in the direction of rolling or perpendicular to it. The coefficient of cubic expansion has been given as $8\cdot9 \times 10^{-5}$ degC^{-1}.

Pure zinc has a resistivity of $5\cdot9 \times 10^{-6}$ Ω.cm at 20°C and the temperature coefficient of resistivity is equal to $4\cdot2 \times 10^{-3}$ degC^{-1} over the range 0—100°C.

For normal commercial zinc the electrical conductivity is around $28\cdot2\%$ that of copper.

MECHANICAL PROPERTIES OF ZINC

Zinc is rather closely related to the low-melting-point group of metals, such as lead, tin and cadmium, and because their softening-point is close to atmospheric temperature it becomes impossible to strain harden them as one would most other metals. Zinc hardens considerably when cold-worked, but one can only take its hot deformation into account when assessing its strength. The main mechanical properties of zinc are as follows:

Minimum yield point for commercially annealed zinc	407 MN/m²
Nominal tensile strength (annealed)	200—500 MN/m²
Elongation for 50-mm samples annealed	27%
Young's modulus of elasticity	190 GN/m²
Approximate annealing temperature	90—150°C

The crystal structure of zinc is shared by magnesium, beryllium and cadmium. Zinc cannot be worked at room temperature but can be rolled at between 150 and 250°C. It becomes brittle when cooled below this temperature range, even slightly lower temperatures having a considerable effect.

The creep of zinc at ordinary temperatures is very high, but can be reduced somewhat by the addition of small quantities of copper, magnesium, titanium, chromium, cobalt or iron. The Rockwell C scale hardness of high-purity cast zinc is about 107 using a $\frac{1}{8}$ in (3·17 mm) diameter ball and a weight of 60 kg for 1 min. The Brinell hardness at 20°C is 12·5 for a 10-mm ball and a load of 500 kg, applied for 30 sec.

The impact resistance of zinc, however, is low. Pressed zinc with a tensile strength of 400 MN/m² has been found to have an impact resistance of only 0·65 kg.m/cm² when using a single blow notched impact test. Alloying of zinc has little effect upon the impact resistance. The shear strength of zinc is about 260 MN/m², or around 30% of that of mild steel.

CHEMICAL AND CORROSION RESISTANCE OF ZINC

Commercial zinc is readily attacked by dilute acids with the evolution of hydrogen, and is also vigorously attacked by nitric acid, although in this case no hydrogen is evolved. It reacts with alkalis to form zincates, though not as actively as aluminium. Zinc reacts readily with sulphur and chlorine in the presence of moisture. When zinc is exposed to normal outdoor atmospheres, it forms a protective coating consisting of zinc oxide, zinc carbonate, basic zinc sulphate, etc. These coatings are usually about 0·007 mm in thickness, are very tenaceous and have the same coefficient of expansion as the metal itself. In addition, these salts are all light-coloured and do not stain other building materials. After the protective coating is formed the rate of corrosion

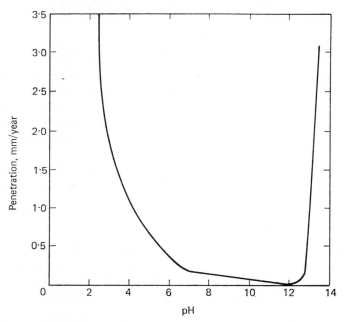

Fig. 7.5   Penetration of zinc versus pH of water in contact with the metal

is uniform and the life of zinc is then proportional to its thickness. The rate of corrosion does, however, vary with the kind of atmosphere in which the zinc is used. For example, the rate of corrosion in moist polluted atmospheres is about ten times as high as in dry and clean air. Zinc in general has a poorer corrosion resistance in air than either lead, nickel or copper. The following rates (g/m² per year) of attack are given by Bayer with respect to zinc:

Pure country air	7—12
City air	20—40
Sea air	17—50
Industrial air	40—80

Zinc is not corroded to any appreciable extent by pure water free from CO_2 and O_2. In the presence of oxygen, zinc hydroxide is produced, which is sparingly soluble, while in the presence of carbon dioxide, zinc carbonate is formed which is rather more soluble. Small quantities of chlorides, sulphates and nitrates in the water markedly increase the attack on the metal. The pH of water in contact with zinc also affects the rate of corrosion very considerably (Fig. 7.5). Corrosion rates are particularly low between pH 6—12·5. When hot water is used, the corrosion rate increases considerably.

Sea water attacks zinc as readily as it does iron and steel with an over-all average corrosion rate of 18 mg/dm² per day with pitting developing to a depth of 0·015 cm. Zinc, however, has considerable resistance to attack by soils, except when they are alkaline with a high chloride content, and acid soils.

Zinc is inert to dry gases and common organic substances such as petrol, paraffin oil or grease. It is attacked by mineral oil at high temperatures and by chlorohydrocarbons in the presence of moisture. The following chemicals also tend to attack zinc: most acids and alkalis, chlorides, nitrates and sulphates, moist cement and ammonium phosphate.

Most compounds of zinc are not too toxic, but the maximum limit of dissolved zinc in drinking water considered safe is 40 mg/litre.

USES OF ZINC

The main uses of zinc connected with the building industry are the following:

As a coating metal
This is the largest single use for zinc in industry and amounts to almost half the total production of zinc.

In the form of rolled zinc
Zinc sheet and strip are used for roofing, gutters, rainwater pipes, flashings and weatherings and, provided the proper techniques are followed, give long maintenance-free service. Zinc alloys containing small additions of copper and titanium have recently been developed and these alloys show better creep strength than unalloyed zinc. They are used for prefabricated roofing, ventilation ducting and pressings.

Diecasting alloys
High-grade zinc with a minimum of 99·99% purity is alloyed with 4% aluminium and 0·04% magnesium and, when a somewhat harder alloy is needed, with 1% copper as well. These die-castings are used in the building industry for such purposes as locks, door handles, bathroom fittings and similar builder's hardware. The properties of diecasting alloy 'A' to BS 1004 containing 4% Al and 0·04% Mg are as follows:

Density, kg/dm³	6·7 kg/dm³
Melting-point	387°C
Solidification shrinkage	1·2%

Linear coefficient of thermal expansion	2.7×10^{-5} degC^{-1}
Thermal conductivity	1·13 kW/m.degC
Tensile strength	283 MN/m²
Elongation for 50-mm samples	15%
Brinell hardness at 20°C with 10-mm ball and load of 500 kg for 30 sec	83 kg/mm²

ZINC COATINGS

Five methods of applying zinc on to base metals are used.

Hot-tip galvanizing

This is applied mainly to semifabricated material such as sheet, strip wire and tube but also increasingly to finished articles. The iron and steel objects are descaled and degreased, followed by immersion in a bath of zinc at 455°C. Various fluxes are used to ensure adherence of the coating. The advantages of hot-dipping are that thick and even coatings are applied, the amount of zinc usually applied varying between 650 and 780 g/m² (Fig. 7.6). All edges, rivets

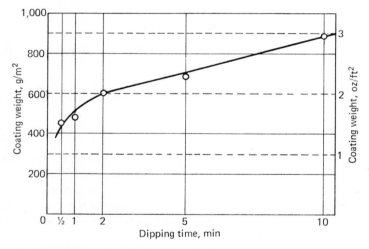

Fig. 7.6   Typical variation of coating weight with dipping time at 455°C during galvanizing processes (By courtesy of the Zinc Development Association)

and seams are duly sealed. In some cases the dipping of articles into the hot zinc is undesirable as it may cause warping and other effects. Other elements are often added to produce special effects. For example, tin and antimony give spangle effects, while aluminium produces a more attractive surface.

Zinc spraying

Atomized particles of molten zinc are sprayed onto the surface, which has been well abraded. The zinc is fed into the pistol in the form of a molten material, a powder or as a wire. Zinc spraying can be applied on site to very large objects. It is a method of applying really thick coatings, which are of the order of 0·25 mm and thicker where needed. Zinc spraying is rather unsatisfactory where coatings are to be applied inside cavities, etc. Open structures such as wire mesh are not economical to treat in this way.

Zinc plating

Coatings of zinc can be applied to iron and steel surfaces by electrochemical means. The steel surfaces have to be prepared very carefully and the objects are then placed in acid sulphate solutions or into alkaline cyanide solutions. The latter is to be preferred when good 'throwing power', i.e. ability to deposit in recesses and hidden parts, is needed. Electrogalvanizing is only capable of depositing very thin zinc coatings, which seldom exceed 0·05 mm in thickness. The advantage of electrogalvanizing is that the article is never heated and that the thickness of the deposit can be controlled within fine limits. It is a process which can be used with irregularly shaped objects.

Sherardizing

This process is used for applying a coating of zinc to objects by heating them for several hours in powdered zinc, just below its melting-point. It is a method that is usually applied to small objects such as nuts and bolts, springs, etc. The maximum size of the objects treated is governed by the size of drum used. The revolving drum is filled with zinc dust together with inert materials such as sand or alumina and the temperature is raised to 380°C. After several hours the objects are separated from the zinc dust and sand by sieving. The objects are usually coated with between 150—300 g/m² of zinc. The surface of the sherardized material consists of a zinc–iron alloy and the process has the advantages that surface hardness is higher than with steel objects treated by other methods and that the thickness of coating is very uniform. Sherardized articles can be painted without requiring any special treatment.

Painting with zinc-rich paints

Paints that leave a dry film containing up to 92—95 % zinc have recently been developed. The medium is usually plasticized polystyrene, and chlorinated or isomerized rubber. The paints can be applied by any of the usual means of spraying, brushing and dipping. They have the advantage of ease of application but the zinc dust does not generally adhere as well as zinc applied by other means. Furthermore the zinc coating is fairly porous and protects the underlying steel only by electrolytic action.

CORROSION RESISTANCE OF ZINC COATING

Under any specific degree of atmospheric pollution the life of a zinc coating is exactly proportional to its thickness. In addition, the life expectancy of a

K

zinc coating depends very much upon the geographical location. In Table 7.3, which is due to J. C. Hudson of the University of Sheffield, the estimated life of a 600 g/m² coating of zinc is given.

TABLE 7.3

Location	Climate	Life, years
Khartoum (Sudan)	Dry and tropical	300
Abisko (Sweden)	Subpolar	300
Aro (Nigeria)	Inland, damp and hot	160
Singapore (Malaysia)	Marine, tropical	100
Llanwrtyd Wells (UK)	Rural, temperate	34
Calshot, Hampshire (UK)	Marine, temperate	23
Woolwich, Kent (UK)	Industrial	19
Motherwell (UK)	Industrial	17
Sheffield (Attercliffe) (UK)	Highly polluted	5

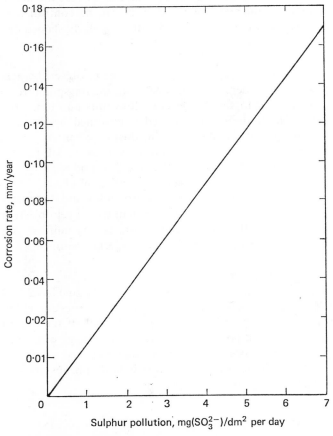

Fig. 7.7 Effect of sulphur pollution on the corrosion rate of zinc

The values in Table 7.3 have been calculated on the assumption that when rusting of the underlying steel begins some 10% of the zinc is still in position. It has also been found that the rate of corrosion of zinc is directly proportional to the degree of sulphur pollution in the atmosphere (Fig. 7.7). The effect of pH on the rate of corrosion of zinc is given in Fig. 7.5.

ACTION OF WATER ON ZINC COATINGS

Softer waters which contain more oxygen and carbon dioxide attack zinc more rapidly than hard waters. Above 60°C zinc becomes cathodic to iron and no longer provides sacrifical protection where the zinc film is broken. It is important to obtain a scaling coat on hot water tanks to protect the steel. Magnesium sacrificial anodes are sometimes used in such conditions.

FINISHES FOR ZINC COATINGS

The simplest finish is chromate passivation treatment by dipping the zinc-coated material into a chromic acid solution. It is particularly advisable when the zinc coating has to withstand humid conditions. Organic finishes can usually be applied to zinc surfaces which have been suitably primed.

Literature Sources and Suggested Further Reading

1. BUTTS, A., *Copper*, Reinhold, New York (1954)
2. COPPER RESEARCH AND DEVELOPMENT ASSOCIATION, Technical Brochures and Information
3. CONSOLIDATED MINING AND SMELTING COMPANY OF CANADA LIMITED, *The Lead Industry*, Trail, British Columbia (1944)
4. CONSOLIDATED MINING AND SMELTING COMPANY OF CANADA LIMITED, *The Zinc Industry*, Trail, British Columbia (1948)
5. LEAD DEVELOPMENT ASSOCIATION, Technical Brochures and Information
6. MATHEWSON, C. H., *Zinc, the Metal, Its Alloys and Compounds*, Reinhold, New York (1959)
7. ZINC DEVELOPMENT ASSOCIATION, Technical Brochures and Information

Addresses: Copper Research and Development Association, 55 South Audley Street, London W1

Lead and Zinc Development Associations, 34 Berkeley Square, London W1

Chapter Eight   Corrosion

Corrosion can be defined as the chemical attack on metals, usually iron and steel, which tends to convert the metal into metallic compounds. This process proceeds spontaneously as it is a process that is exothermic. However, the processes are far more than simple chemical reactions, but are basically electrochemical.

The corrosion processes can be explained in the simplest way by considering a standard torch battery (see Fig. 8.1). In this the zinc anode is in contact

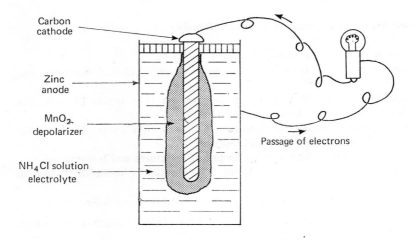

Anode reaction at zinc surface

$$Zn \rightarrow Zn^{2+} + 2e$$

$$Zn^{2+} + 2Cl^- \rightarrow ZnCl_2$$

Cathode reaction at carbon rod

$$2H^+ + 2MnO_2 + 2e \rightarrow H_2O + Mn_2O_3$$

Fig. 8.1 Torch battery

with a solution of electrolyte, namely ammonium chloride. At the centre of the cell there is a carbon cathode, surrounded by an oxidizing agent, manganese dioxide. Normally, when no electrical contact exists between the carbon and the zinc, the zinc remains unattacked. When, however, a wire is connected from anode to cathode, zinc goes into solution in the form of an ion:

$$Zn \rightarrow Zn^{2+} + 2e$$

The electrons flow to the carbon cathode via the wire. Hydrogen ions from the electrolyte migrate to the cathode too, and react with the manganese dioxide depolarizer and the electrons which have passed to the carbon cathode as follows:

$$2MnO_2 + 2H^+ + 2e \rightarrow Mn_2O_3 + H_2O$$

If the carbon were not surrounded by the depolarizer, the reaction would also have come to a stop, as the reaction:

$$2H^+ + 2e \rightarrow H_2$$

cannot take place in neutral solutions because of the so-called 'polarization' and 'hydrogen overvoltage' effects. The first is the back-e.m.f. induced when a current is flowing, while the second is the difference in potential between a cathode at which hydrogen is evolved and a hydrogen electrode at equilibrium in the same solution. This overvoltage depends upon the nature of the cathode surface, varying between about 0·03 V for platinium and up to 0·20 V for cadmium. For most metals the overvoltage amounts to 0·10—0·12 V.

The calculation of polarization values is somewhat more complicated and depends on the concentration of the solution, the current passing, as well as the nature of the cathode and the ions concerned.

The driving force which makes the torch battery reaction go is the difference in natural e.m.f. between the two materials, the zinc anode and the carbon cathode.

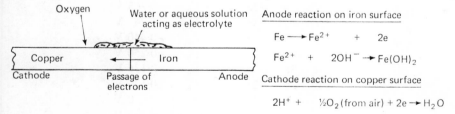

Fig. 8.2   Corrosion cell

Corrosion of iron and steel, or indeed a non-ferrous metal, proceeds in an exactly analogous fashion. To start off with, any difference in natural e.m.f. of the metals determines which is going to be the anode and which the cathode. When steel is in contact with a metal which is naturally cathodic to it, such as copper, its role is immediately determined. But even two samples of steel, or even two adjoining crystals of steel, can have different natural e.m.f's and in this case one sample or crystal becomes the anode, which corrodes, while the other becomes the cathode, whose function is to dispose of the hydrogen ions liberated. The place of the manganese dioxide depolarizer is taken by atmospheric air, which oxidizes the hydrogen ions together with free electrons into water. In fact, the relative abundance of oxygen also can decide which sample of steel or even which crystal of steel becomes the cathode and which the anode. A high degree of oxygenation tends to make the sample of steel underneath cathodic.

The complete corrosion reactions can now be summarized as follows:

$$At\ anode \quad Fe \rightarrow Fe^{2+} + 2e$$
$$Fe^{2+} + 2OH^- \rightarrow Fe(OH)_2$$

The ferrous hydroxide then changes slowly under the action of various atmospheric agencies into a loose porous brown mass which contains the following compounds in various proportions: $Fe(OH)_2$, $Fe(OH)_3$, FeO, Fe_2O_3, Fe_3O_4, $FeCO_3$, $Fe_2(CO_3)_3$, $FeSO_4$, FeS, etc.

$$At\ cathode \quad 2H^+ + 2e + \tfrac{1}{2}O_2 \text{ (from air)} \rightarrow H_2O$$

8.1 The galvanic series of natural e.m.f's of metals

To determine which of the two metals, in a pair, is likely to become the anode and corrode, and which is likely to remain the cathode, one makes use of the standard electromotive series. This gives the e.m.f. in volts at 25°C, in relation to hydrogen. The most common pure metals dealt with in the building industry are as given below:

$$Mg \rightarrow Mg^{2+} + 2e \qquad E_0 = +2 \cdot 37 \quad V$$
$$Al \rightarrow Al^{3+} + 3e \qquad E_0 = +1 \cdot 66 \quad V$$
$$Ti \rightarrow Ti^{2+} + 2e \qquad E_0 = +1 \cdot 63 \quad V$$
$$Zn \rightarrow Zn^{2+} + 2e \qquad E_0 = +0 \cdot 763 \ V$$
$$Cr \rightarrow Cr^{3+} + 3e \qquad E_0 = +0 \cdot 74 \quad V$$
$$Fe \rightarrow Fe^{2+} + 2e \qquad E_0 = +0 \cdot 440 \ V$$
$$Ni \rightarrow Ni^{2+} + 2e \qquad E_0 = +0 \cdot 250 \ V$$
$$Sn \rightarrow Sn^{2+} + 2e \qquad E_0 = +0 \cdot 136 \ V$$
$$Pb \rightarrow Pb^{2+} + 2e \qquad E_0 = +0 \cdot 126 \ V$$
$$H^2 \rightarrow 2H^+ + 2e \qquad E_0 = \qquad 0 \ V$$
$$Cu \rightarrow Cu^{2+} + 2e \qquad E_0 = -0 \cdot 337 \ V$$

In any galvanic cell, the e.m.f. between the two electrodes is given by the following equation:

$$E = \left[E_0\,(A) - \frac{0 \cdot 0592}{n_A} \log_{10} \text{(ionic concentration of } A \text{ ions)} \right]$$
$$- \left[E_0\,(B) - \frac{0 \cdot 0592}{n_B} \log_{10} \text{(ionic concentration of } B \text{ ions)} \right]$$

where $E_0(A)$ and $E_0(B)$ are the standard electrode potentials of A and B and n_A and n_B are the valencies of A and B. If E is a positive value, then metal A becomes the anode and corrodes, while metal B becomes the cathode and is unattacked. If E, however, is a negative value, metal B becomes the anode and corrodes, while metal A is the cathode. Everything else being equal, a high value of E means that corrosion is likely to be rapid.

Metals that are particularly harmful to iron and steel when in contact with it are those which are not only possessed of a very low standard oxidation potential, but also are virtually insoluble in the usual corrosive agents surrounding the iron–metal couple. In consequence one of the most harmful is copper.

It is also possible for metals or alloys that have intrinsically a higher standard oxidation potential than iron to be cathodic when in contact with the iron. Aluminium and chromium are cases in point. They get covered by passive films, so that the metal is never exposed to the electrolyte at all, and in consequence both these metals behave like oxygen electrodes. Instead of becoming anodes, as their position in the electrochemical series would suggest, they often, in actual fact, behave like cathodes, accelerating the corrosion of iron and steel in contact with these metals.

On the other hand, when tinned steel is in contact with food on the inside of tin cans, the formation of tin complexes with food acids actually makes the tin anodic with respect to steel, thereby protecting the latter. This does not happen when tin plate is exposed to the atmosphere. Under such circumstances the tin is cathodic, causing rapid corrosion of the steel, if the steel should at any time be exposed to the atmosphere due to a scratch, etc.

8.2 Practical electrochemical series

Owing to such effects as passivity, polarization, etc., the electrochemical nature of metals varies with the type of electrolyte to which they are exposed. The list below gives the relative position of common commercial alloys when in contact with sea water. This gives a rough indication too regarding their relative position in normal industrial atmospheres. The list is written with the most anodic at the top, and one can therefore assume that in most cases, where couples are made up of two members of the list, the one closer to the top of it will corrode while the other will act as the cathode. This is, however, by no means true in all cases and this list should be taken as an approximate guide only.

> Magnesium, magnesium alloys, zinc, pure aluminium–cadmium, duralumin, mild steel, wrought iron, cast iron, 50/50 lead–tin solder, 18/8 stainless steel (active), lead, tin, bronzes, nickel (active), Iconel (Ni/Cr/Fe alloy) (active), brasses, copper, silicon bronze, nickel (passive), Iconel (passive), 18/8 stainless steel (passive), Cr/Ni/Mo stainless steel (passive).

As can be seen, the existence of a passive (usually a coherent film of oxides, etc.) coating on the surface of the metals makes an enormous difference with regard to their position in the electropotential series. This means, of course, that while fresh aluminium which is fairly active will still be in the anodic position to iron and steel, the same is not true of aluminium where a good passive film has formed. In the same way, stainless steel, which is still active, will corrode when in contact with copper, bronzes and brasses. When stainless steel has been exposed to the atmosphere, it becomes one of the most cathodic metals in common use.

8.3 Rate of corrosion

The rate of corrosion is determined by the slowest reaction rate of the series of reactions given above. This is often the reaction:

$$2H^+ + 2e + \tfrac{1}{2}O_2 \rightarrow H_2O$$

which takes place on the cathodic side of the cell. This is conditioned by the following factors:

1 Ready availability of oxygen or other oxidizing agents.
2 Relatively large surface area of cathode.
3 Freedom from encrustation of cathodic surface.

There are, however, other cases where the rate-determining step is the anodic reaction. In such a case the corrosion rate depends on the following:

1 Difference in effective e.m.f. between anode and cathode.
2 Ready conductance away of electrons from anode to cathode.
3 Ease of access of electrolyte to anodic surface.
4 Degree of ionization of electrolyte, i.e. presence of quantities of ionizable salts.

The rate of corrosion of iron and steel is particularly high when the pH of the electrolyte is below about 3 (see Fig. 8.3). Under such circumstances polarization effects no longer stop the liberation of hydrogen gas from the cathode and in consequence the rate of corrosion steps up enormously. On the other hand, when the pH of the electrolyte is high, iron is covered with a passive film which effectively prevents corrosion.

8.4 Forms of corrosion affecting steel structures

DIFFERENTIAL METAL CORROSION

When steel is in contact with a metal which is below it in the electrochemical series it becomes very liable to corrosion. The metals that are particularly harmful to steel are copper, the bronzes, brasses, lead, tin, chromium and passive aluminium. The reason is that in such circumstances steel becomes strongly anodic, and is liable to be attacked under even quite mildly corrosive

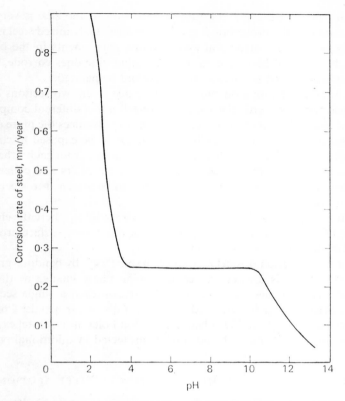

Fig. 8.3 Effect of pH on corrosion of steel in aerated water at 25°C

conditions. Corrosion conditions are at their worst when the surface of the steel exposed is small, while that of the cathodic metal is large. For example, steel rivets holding copper plates together would corrode very rapidly indeed, as would steel nuts and bolts in contact with large surfaces of anodized aluminium. Such conditions must always be avoided. On the other hand, one can use copper or brass rivets, bolts and other small fasteners together with large surfaces of steel without appreciably increasing the risks of corrosion. This is also the reason why, when one has to join copper and steel plates, it is best to leave the steel pipes unpainted, but one should coat the copper pipe with as thick a layer of bitumen, or other coating, as possible.

When steel is coated with zinc, the steel becomes the cathode and the zinc the anode. Corrosion of galvanized steel sheeting takes place very slowly indeed, as the cathodic surface of the steel is small, being only in the form of scratched sections or pinholes penetrating the zinc coating. Furthermore, the zinc becomes somewhat passive, but not passive enough to counteract the difference of natural e.m.f. between it and the steel, as is the case with aluminium. This means that the rate of corrosion of the zinc is very slow, without impairing the protection it gives to the steel. Galvanized steel is therefore justly popular for structural and cladding components in the building industry. It must not be forgotten, however, that zinc does corrode, even if slowly, and galvanized surfaces should therefore be painted.

Differential metal corrosion can also take place when two sections of steel are fastened together, particularly if the two steels are of different composition (mild steel against medium steel, etc.). In such circumstances the more electropositive section will corrode rapidly. Joints likely to be exposed to corrosive conditions should thus always be electrically insulated from each other. This may be done by using thin plastic packing pieces, washers and collars. This also applies when, say, aluminium flashing has to be attached to steel framework sections.

The rate of corrosion is always highest close to the joint, where the electrical resistance to the corrosion currents is least. For this reason surface protection of the joint should be carried out meticulously.

When a single section of steel is strained in some way, by bending, grinding, drilling, etc., there is always an accompanying phase change at the place where the metal has been worked. For this reason even a single section of steel then behaves as if it were made up of two dissimilar metals. Corrosion takes place particularly readily at bends, near bolt holes, near scratches and cut edges, etc. Such parts should be adequately protected by additional painting.

ATMOSPHERIC CORROSION OF A SINGLE STEEL SECTION

On single steel surfaces corrosion frequently takes place at grain boundaries. A usual initial cause is a state called differential aeration. If a section of a steel surface is more exposed to the atmosphere than a neighbouring section,

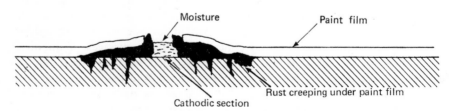

Fig. 8.4   Corrosion at point of damage of a paint film

it becomes cathodic. The rust formed ensures that the coating material, whether in the form of paint, plastic coating, stove enamel or vitreous enamel, is lifted off the surface, thereby exposing further areas of metal and causing the corrosion to creep along underneath the coated material (see Fig. 8.4).

The rate of corrosion which takes place with imperfectly surface-coated metals is often far more rapid and more destructive than it would be with metals which are left uncoated. The reason for this is that the anodic action, which is otherwise spread out along the whole surface, is concentrated at particular spots, which are pitted so deeply that they may be punctured (see Fig. 8.5). Once a surface is rust-encrusted the rate of further corrosion tends to fall appreciably (Fig. 8.6).

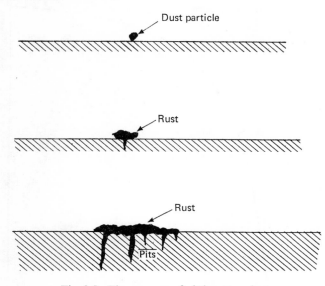

Fig. 8.5 The progress of pitting corrosion

The rate of corrosion is markedly affected by the cleanliness of the atmosphere or otherwise. Hot and dry atmospheres produce a low degree of corrosion, while even cold and damp atmospheres do not cause over-rapid corrosion provided there are few dissolved acid gases. Under damp and steamy conditions the rate of corrosion rises, particularly if the area is near the sea. But by far the worst conditions affecting corrosion are atmospheres loaded with carbon particles and sulphur dioxide as exist in industrial areas.

TABLE 8.1

Location	Climate	Corrosion, $mm \times 10^{-3}$/year
Khartoum (Sudan)	Dry and tropical	0·71
Abisko (Sweden)	Cold and damp, clean atmosphere	2·50
Aro (Nigeria)	Inland, damp and hot	4·08
Singapore (Malaysia)	Hot and damp, marine atmosphere	13·3
Llanwrtyd Wells (UK)	Small town, rural, temperate	47·5
Calshot, Hampshire (UK)	Residential, marine atmosphere, temperate	53·5
Motherwell (UK)	Industrial atmosphere	80
Woolwich, Kent (UK)	Industrial atmosphere	88
Sheffield (Attercliffe) (UK)	Highly polluted industrial atmosphere	99

In a classical experiment, J. C. Hudson exposed ingots of iron to various atmospheres around the world and measured the degree of corrosion after one year in terms of thousandths of a millimetre. His findings are given in Table 8.1.

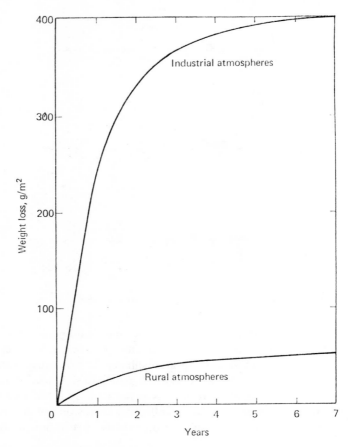

Fig. 8.6   Rates of corrosion of mild alloy steel and protective action of rust encrustation

CORROSION CAUSED BY ELECTRIC CURRENTS

Large sections of steelwork, buried pipelines, etc., often carry large electric currents. These arise due to leakages from power and telephone cables, thunderstorms and in other ways. It has been found that buried water pipes may at times carry as much as 50 A. In the building industry stray current corrosion (see Fig. 8.8) has been found with respect to steel stanchions encased in concrete foundation pads, which were cast using additions of calcium chloride as antifreeze. If the current is discharged from the steel

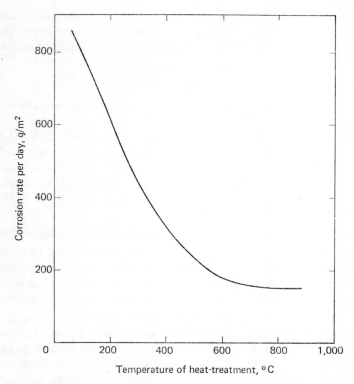

Fig. 8.7   Effect of heat-treatment on the rate of corrosion of cold-worked steel with 0·1 % C in 0·1 N hydrochloric acid at 25°C

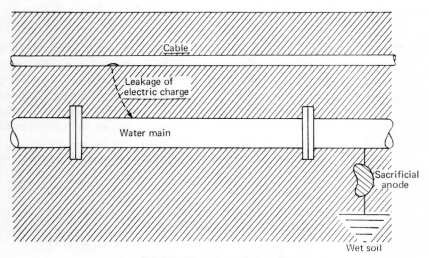

Fig. 8.8   Stray current corrosion

section into earth at a position where the soil is damp, the anodic action of the corrosion process takes place, and extremely rapid corrosion of the steel is the result. The reactions which take place can be summarized as follows:

$$Fe \rightarrow Fe^{2+} + 2e \text{ (to earth)}$$
$$Fe^{2+} + 2OH^- \rightarrow Fe(OH)_2$$

producing all the usual rust end products in the presence of various acidic oxides and solutions.

The rate of corrosion which takes place does not depend upon the rate of the cathodic process but on the quantity of current that leaves the conductor. Owing to the fact that the currents are so very heavy and that the anodic and cathodic products are produced some distance apart and are unlikely to eliminate each other by interaction, the rate of corrosion tends to be high. Three methods of protection can be used:

1 *Insulation of joints* reduces the amount of stray current which can build up.

2 *Electrical drainage* consists of connecting the structure to a positive terminal at a place where no corrosive conditions occur so that the stray current exits there rather than to earth. 'Overdraining' considerably reduces the likelihood of the structure being damaged by corrosion.

3 *The use of sacrificial anodes*. Large lumps of pig iron, or even more frequently, lumps of zinc are connected to the steel structure at particularly corrosive positions, and the current is encouraged to exit to earth at these positions. Naturally heavy corrosion takes place in these sacrifical anodes, but the main steel structure is thereby protected.

Painting is usually worse than useless; it merely concentrates corrosion at the weakest section.

ANAEROBIC CORROSION

This form of corrosion differs from others in that it proceeds in the absence of air or oxygen. In fact, anaerobic corrosion can be stopped most effectively by the supply of oxygen to the affected parts. Anaerobic corrosion takes place mainly in buried steel pipes and similar objects. It is due to bacterial action, the most important of the bacteria causing such corrosion actions to take place being one called *Vibrio desulphuricans*, which catalyses the reduction of sulphates in the soil.

The reactions which take place under conditions of anaerobic corrosion are the following:

At anode $\quad Fe \rightarrow Fe^{2+} + 2e$
$$Fe^{2+} + 2OH^- \rightarrow Fe(OH)_2$$
$$Fe(OH)_2 + CaS \rightarrow FeS + Ca(OH)_2$$
$$Ca(OH)_2 + H_2SO_4 \text{ (from soil)} \rightarrow CaSO_4 + 2H_2O$$
At cathode $\quad CaSO_4 + 8H^+ + 8e \rightarrow CaS + 4H_2O$

The calcium sulphide then drifts over to the anodic side and completes the anodic reaction. The cathodic reaction is the one which is catalysed bacterially. When the reaction cycle is finished the steel has been replaced by a mixture of ferrous compounds, $Fe(OH)_2$:FeS in the ratio of about 3:1, which has no protecting power, as it is in the form of a very loose deposit.

The degree of attack by this type of corrosion is markedly dependent upon the nature of the soil. It has been found that light sandy soils are almost without effect, as are soils which are alkaline. The maximum amount of corrosion will be found in waterlogged acid soils (peats and clays especially). Anaerobic corrosion once started tends to accelerate due to the multiplication of the bacteria.

Anaerobic corrosion can be stopped by the addition of small quantities of selenium and tellurium to the steel, which effectively inhibit this kind of corrosion. Alternatively, one can encase the pipes or steel sections in thick coatings of bitumen.

CORROSION IN WATER AND OIL TANKS

In water tanks the rate of corrosion is a maximum just underneath the water-line (see Fig. 8.9). The tank wall above the meniscus functions as the cathode, and serves to oxidize hydrogen ions and the electrons produced during the anodic action to water. The anodic action takes place underneath the water surface, at a position closest to the surface.

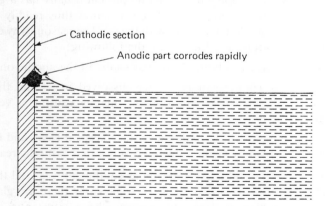

Fig. 8.9 Waterline corrosion

Water tanks should be heavily galvanized, as zinc is not readily attacked by hard water.

In the case of oil tanks, oil pipes and other equipment of the same type, there is a very considerable danger of corrosion if water is allowed to accumulate at the bottom (see Fig. 8.10). This water is usually heavily polluted by

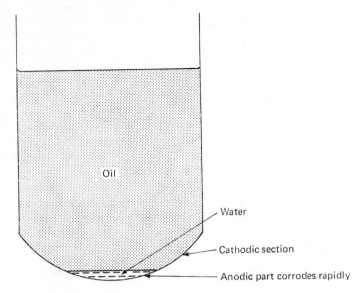

Fig. 8.10   Corrosion of oil tanks

chlorides and sulphates extracted from the oil, and readily induces the anodic reaction on the metal section adjacent to it. The electrons and hydrogen ions flow upwards, towards the metal covered by oil. Oil usually has a good deal of oxygen dissolved in it, and the cathodic reaction thus readily proceeds against the metal surface left in contact with the oil. The reasons why corrosion is so rapid at the bottom of oil tanks are the following:

1 The cathodic surface is large in comparison with the anodic one.

2 The oil usually contains more oxygen dissolved in it than the partial pressure of oxygen in the atmosphere, thereby causing a very ready depolarizing action.

3 The oil in contact with the cathodic surface keeps it clean and free from encrustations, thereby again inducing a very rapid cathodic reaction.

To prevent corrosion of oil tanks and similar equipment it is vital that water is drawn off as soon as it is deposited. In some cases there are glass water traps fitted, and in others, an alarm system warns whenever water has deposited at the bottom of the oil.

STRESS CORROSION

It has been found that a sample of cold-worked steel corrodes many times as fast as heat-treated steel when exposed to an acid environment. The reason for this is the formation when cold-working the steel of a crystalline structure containing adjoining particles with a different natural e.m.f. Under conditions of heat-treatment, a more uniform solid solution is formed (see Chapter 6).

Mild steel has been found to crack much more readily when exposed in the stressed state to corrosive conditions than normally (see Fig. 8.11). It is found to fracture along intergranular paths. Such cracking is called stress corrosion. Practical examples are the cracking near, and of, rivets in riveted boiler plants

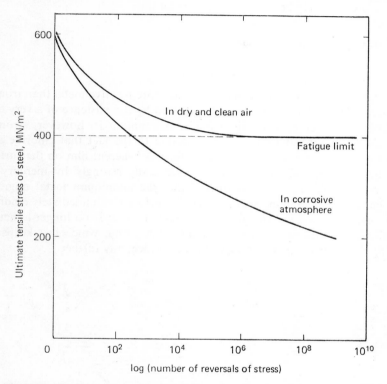

Fig. 8.11   Behaviour of steel under cyclic stresses in air and corrosive conditions

and similar equipment, steel cables exposed to nitrates, and stressed steel tanks containing liquid ammonia and other corrosive substances. In fact, whenever stressed steel objects are kept for any length of time in corrosive atmospheres or in contact with even traces of corrosive liquids and solutions, care must be taken to avoid the often disastrous effects of corrosion cracking.

The precautions taken to avoid stress corrosion are the following:

1 *Heat-treatment.* Steel cooled slowly from about 900—950°C or quenched and then tempered at 250°C is fairly immune to corrosion cracking.

2 *Alloying.* Small quantities of Al, Ti, Nb and Ta very much reduce stress cracking, although results are not 100%.

L

3 *Surface treatment of steel.* By shot-blasting the steel, it has been found that the surface becomes stressed and protects the body of the steel against stress cracking.

Closely allied to stress corrosion is corrosion fatigue. Whereas steel that is subjected to cyclic stresses in clean conditions stabilizes at a fairly high fatigue limit, steel exposed to corrosive conditions and subjected to cyclic stresses progressively becomes weaker until it fails.

8.5 Corrosion of non-ferrous metals

ALUMINIUM

Although aluminium is fundamentally a far more reactive metal than iron, it is protected exceedingly well against corrosion by the existence of a very thin and coherent layer of aluminium oxide. Aluminium is, however, strongly attacked by many salt solutions. This is due to the fact that chloride and sulphate deposits, unlike oxides, do not form a coherent film on the surface of aluminium. Aluminium is attacked especially strongly by mercury or mercury salts. An amalgam is formed and the aluminium metal migrates through the amalgam film. As it reaches the surface it is immediately oxidized by atmospheric oxygen. The aluminium oxide formed is no longer adherent and therefore does not protect the aluminium metal, which just keeps on drifting through the amalgam layer to the surface, any further.

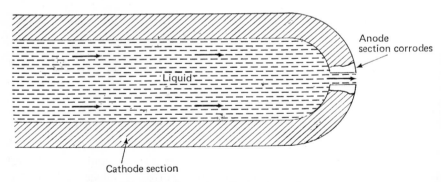

Fig. 8.12 Corrosion of copper nozzles

COPPER

In the absence of moisture, copper is hardly attacked at all, but under normal conditions copper is covered with an adherent protective film, which is a mixture of $CuCO_3$, $Cu(OH)_2$, $CuCl_2$ and $CuSO_4$.

When liquids containing dissolved salts flow past copper, corrosion takes place at the part where liquid motion is at its most rapid, which becomes the

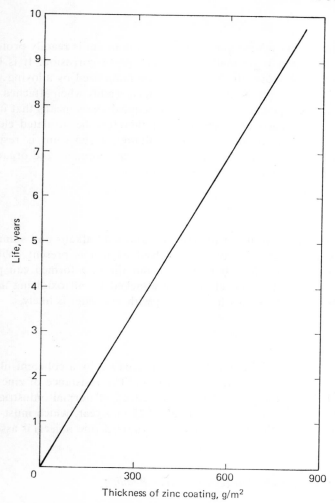

Fig. 8.13 Relationship between thickness of zinc coating and
life expectancy at Sheffield (after Hudson)

anodic section. Unlike iron, no encrustation is formed and the copper simply
appears to wear away. This phenomenon is found with copper nozzles used
in central heating systems and also with some plumbing fittings (see Fig. 8.12).

LEAD

Lead is usually well protected by a surface film, but becomes liable to corrosion
when buried in loams and other acidic soils. Lead pipes are also liable to
corrosion by stray currents, which affect them in the same way as steel pipes.

MAGNESIUM

Magnesium is very active indeed, yet like aluminium is readily protected by an oxide film, and is normally fairly immune to corrosion. It is liable to stress cracking in moist air, but this can be minimized by alloying with 3% Zn and 0.7% Zr. Magnesium can be corroded readily when attached to other metals as its high position in the electrochemical series means that it will act as an anode in nearly all cases. It must therefore be insulated electrically from all dissimilar metals. Unlike aluminium, magnesium is resistant to alkalis. It is attacked most readily by sea water, inorganic and organic acids and acid salts, methanol and leaded petrols.

NICKEL

Nickel is extremely resistant to both hot and cold alkalis and somewhat to dilute non-oxidizing acids unless dissolved oxygen is present. With many aerated aqueous solutions it is passive, but the film formed can pit when exposed to sea water. Nickel is easily attacked by all oxidizing acids and should also never be used where contact with sea water is likely.

ZINC

Between pH limits of 6.0 and 12.5 zinc is covered by a coherent film which protects it extremely well against corrosion. The resistance of zinc to acids and alkalis is not too good. The rate of attack of normal industrial atmospheres (see Fig. 8.13) on zinc is about 0.025 mm/year, which must be taken into account when the life duration of galvanized zinc objects is assessed.

Literature Sources and Suggested Further Reading

1. APPLEGATE, L. M., *Cathodic Protection*, McGraw-Hill, New York (1960)
2. BREGMAN, J. J., *Corrosion Inhibitors*, MacMillan, London (1963)
3. CHAMPION, F. A., *Corrosion Testing Procedures*, Chapman and Hall, London (1964)
4. EVANS, U. R., *The Corrosion and Oxidation of Metals*, Arnold, London (1950)
5. MANN, J. Y., *Fatigue of Materials*, Melbourne University Press, Melbourne (1967)
6. MORGAN, J. H., *Cathodic Protection*, Leonard Hill, London (1959)
7. QUE, F. L. LA, and COPSON, H. R., *Corrosion Resistance of Metals and Alloys*, Reinhold, New York (1963)
8. PUTILAVA, I. N., BALEZIN, S. A., and BARANNIK, V. P., *Metallic Corrosion Inhibitors*, Pergamon Press, Oxford (1960)
9. SHREIR, L. L., *Corrosion*, 2 volumes, Newnes, London (1963)

10. ROBERTSON, W. D., *Stress Corrosion Cracking and Embrittlement*, Wiley, New York (1956)

11. SPELLER, F. N., *Corrosion Causes and Prevention*, McGraw-Hill, New York (1951)

12. TATTON, W. H., and DREW, E. W., *Industrial Paint Application*, Newnes, London (1964)

Chapter Nine   Surface Coatings

There are numerous surface coating materials which are used for protective and decorative purposes in the building industry. Surface coatings have the following constituents:

1 *Vehicle or medium.* This is a clear varnish or gummy base, which attaches itself to the surface of the substance to be coated and solidifies there. Some vehicles dry simply by the evaporation of the solvent, but most harden by some chemical action such as polymerization.

2 *Solvent.* The purpose of the solvent is to provide the surface coating material in a form in which it can readily be applied by brushing, spraying or dipping. The solvent is designed to evaporate after the surface coating material has been applied. Solvents vary in nature from water to such organic materials as esters, hydrocarbons, terpenes, ketones, etc.

3 *Pigments.* There are numerous inorganic and organic pigments, all of which are in the nature of insoluble white or coloured materials, chosen for their opacity. Pigments vary in their chemical properties, however, and must be carefully selected from the point of view of inertness to various surfaces and atmospheres.

4 *Extenders.* Extenders are substances that are added in order to increase the 'body' of paints and thus to improve the abrasion and other mechanical resistances. Extenders are also used in matt paints. They are not normally self-colouring.

5 *Driers.* Driers are catalysts which are often added to paints and varnishes in order to increase the rate of setting and to improve the final strength.

6 *Dyes.* When surface coatings are intended to be completely transparent but to give a coloured tint, dyes are used instead of pigments. Surface coatings which incorporate dyes are not, however, very common.

When an organic surface coating contains a vehicle and solvent only, it is termed a 'varnish'. Paints are varnishes which have pigments and/or extenders dispersed.

9.1 Vehicles or media

Vehicles fall into two classes, non-convertible materials and convertible coatings. The *non-convertible* type is one which simply dries on the surface, after the solvent has evaporated off. The most common of these are: shellac, Manila gum, bitumen, waxes, cellulose acetate, non-drying alkyd resins, cyclized and chlorinated rubber, vinyl resins and acrylic resins.

In the case of the *convertible* materials, once the substance has been deposited on the surface of the object in question, its nature changes. Either under the action of atmospheric oxygen, or by the action of catalysis or heat, the molecules join up to produce large and tough polymeric films. The simplest of such materials are the various drying oils, which are unsaturated fatty esters. Typical examples of such drying oils are linseed oil, tung oil, oiticica oil, dehydrated castor oil and soya bean oil, as well as oils made by blending and heat treatment of raw oils.

The second group comprises a whole series of natural and synthetic resins, which can either be used neat, or reacted together with a drying oil in order to produce a more adherent and flexible substance. The main resins used in the surface coating field are rosin, coumarone resin, zinc rosinate, calcium rosinate, copal resin, maleic acid – drying oil compounds, alkyd resins and alkyd resin – styrene copolymers, urea formaldehyde, melamine formaldehyde, phenol formaldehyde and its copolymers, epoxide esters and polyurethanes.

Most surface coatings use a blend of synthetic resins in order to obtain the best results for each specific case.

9.2 Drying oils

Practically all common vegetable oils are mixtures of the triglycerides of fatty acids. But only if these fatty acids have one or more double bonds within their structure can the oil dry by oxidation and consequent film formation. The quantity of oxygen used for this setting process is found to be about 30—40% of the weight of the oil. It is very much accelerated by the presence of so-called 'driers'. For example, if 0·05% of cobalt in the form of cobalt naphthenate is added to linseed oil, it sets in 20% of the time needed for the drying of the neat oil. The rate of drying is also governed by the prevailing temperature, the warmer the conditions the quicker a paint film dries.

To produce a drying oil that is somewhat thicker than the raw oils, one uses a technique called thermal polymerization. If linseed oil is heated for a number of hours at about 285°C it thickens appreciably and is then called 'stand oil'. Blown oils are made by blowing air through the raw oils at a temperature of about 120°C. This also thickens the oil, but the hardened blown oils are not as durable and will tend to yellow more quickly than equivalent stand oils.

9.3 Oleoresinous varnishes

These are clear film-forming liquids which dry by ordinary atmospheric oxidation to form a hard and glossy film. All these varnishes contain drying oils, one or more resins, thinners and driers. The function of the oil is to give the varnish durability on exposure, and flexibility. The resins used are hard and oil soluble, and are dispersed into the oil by heating. Their function is to improve the hardness and the gloss of the varnish. If a suitable resin is used, faster drying can be obtained as well.

The main thinners are white spirit, dipentene, coal tar naphtha, xylol and various petroleum spirits.

Small quantities of wood turpentine are often also added to give the paint a pleasant smell. The most common thinner of all is white spirit. Heavier petroleum distillates such as kerosene retard the rate of drying but improve the flow. Rather more volatile solvents such as xylol and petroleum distillate are used where fast drying is essential. The quantity of solvent used depends upon the viscosity required. Brushing lacquers usually have a viscosity of 2 P at 25°C.

9.4 Driers

The most common driers employed are naphthenates, linoleates and rosinates of lead, cobalt, calcium and manganese (see Fig. 9.1). Lead driers induce polymerization in the body of the film, while cobalt tends to induce primarily surface skinning. A wrong drier balance causes uneven drying of the paint film and may be the reason for surface wrinkling or crazing. Satisfactory drier ratios have been found to be: 0·05% Co, 0·5% Pb and 0·25% Ca for long oil alkyd varnishes and 0·05% Co, 0·025% Mn and 0·25% Ca for medium length linseed oil–tung oil–alkyd clear varnishes. Each type of varnish requires its own specific drier balance, which is usually arrived at by methods of trial and error.

9.5 Classification of varnishes

An important criterion of varnishes is the resin to oil ratio, after the solvents have been driven off. The usual terms used are:

        Resin to oil   1:½ to 1:2   (short oil varnish)
        Resin to oil   1:2 to 1:3   (medium oil varnish)
        Resin to oil   1:3 to 1:5   (long oil varnish)

Short oil varnishes dry quickly to form a hard film with a high gloss, but do not have very good flexibility. They are mainly used for interior purposes in the building industry.

Long oil varnishes dry slowly, but have excellent flexibility and weather well. They are the main type used for exterior painting jobs.

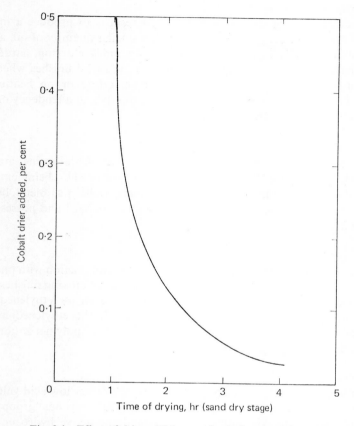

Fig. 9.1   Effect of drier addition on the drying time of a paint

SYNTHETIC RESIN VARNISHES

Alkyd resins

Alkyd resins are made by reacting polyhydric alcohols such as glycerol with dibasic acids such as phthalic anhydride at a temperature of about 220°C. The reaction is stopped before complete cross-linking of the molecules has taken place. Such alkyd resins are, however, never used on their own but are modified by reaction with linseed oil. These linseed-oil-modified alkyd resins are nowadays the most widely used of all in the paint industry and constitute more than half of all resins used.

When it is desired to use light pigments in the paint, it is necessary to employ soya bean oil or sunflower oil instead of linseed oil or China wood oil, as the latter tend to yellow with age.

Alkyd–Styrene copolymer resins

The effect of reacting styrene with an alkyd resin is to produce a material which is capable of being incorporated with a smaller quantity of oil, and yet compatible with solvents. These copolymers are quicker drying, harder, and also impart a pale colour. They are found in industrial finishes where pale colour and rapid drying, either at room temperature or on heating, are needed. Disadvantages are a certain lack in flexibility and a tendency of being readily attacked by hydrocarbon solvents.

Amino resins

Urea-formaldehyde resins and melamine–formaldehyde resins are used almost entirely for application as industrial stoving finishes, being employed for coating the outside of refrigerator cabinets, radiators, metal building components, etc. These resins are cured speedily, are hard and possess good colour retention.

Epoxide resins

Epoxide resins are rather expensive and are used in conjunction with phenolics and urea-formaldehyde plastics for the manufacture of stoving finishes which have good flexibility. Cold-cured epoxide resins, which use ethylenediamine as an initiator, are employed for floor finishes. If the ethylenediamine is mixed with the epoxide finish just before application, polymerization takes place rapidly at room temperature.

Polyurethanes

Polyurethanes are compounds of polyesters with a rather low acid value (less than 10 mg KOH/g) with tolylene di-isocyanate. The isocyanate group is very active and reacts quickly with materials containing active hydrogen atoms such as castor oil, epoxy resins and many other plastics. This reaction is dependent upon temperature and the nature of reacting groups. Polyurethane coatings dry and reach the fully cured stage at room temperature, with the exception of the isocyanate products which are compounds of the isocyanates with materials such as phenol or acetoacetic ester. These cannot set at room temperature, but if they are heated above 145°C, they split up and free isocyanate is released. This then immediately reacts with the polyester resin to form a hard and fully cured material.

When formulating urethane lacquers it is necessary to choose solvents that do not contain free hydroxyl or similar groups as they can react with the isocyanates. The main solvents employed are ketones, butyl and amyl acetates, with toluene, xylene and naphtha as diluents. Urethane surface coatings have poor flow properties so that it is necessary to use flow-out agents such as ethyl cellulose or cellulose acetobutyrate in the formulations. Pigments chosen in connection with urethane finishes should be dry, as water reacts with isocyanates.

Polyurethane finishes have excellent chemical resistance, particularly to oils, solvents and ozone. They can be used satisfactorily up to 150°C. Polyurethane finishes are widely used as timber finishes because of their chemical stability and as floor finishes because of their excellent abrasion resistance. They are also excellent rubber lacquers and are used in this capacity to protect rubber from the effect of ozone in the atmosphere.

Silicones

Silicone resins are usually supplied as a 50—80% solution in aromatic solvents. They are thermosetting resins and need curing at about 230°C. After they have set, silicone finishes can be used at temperatures up to 300°C. Paints based on silicones are virtually unaffected by weathering, and are not attacked by most dilute acids and bases and many other chemicals. On the other hand, silicone finishes do not adhere too well, lack toughness and are not very solvent-resistant. Silicone paints are used in the building industry for painting steel chimney stacks and to provide water-repellant coatings for masonry.

Other resins used

Apart from phenolic resins, which are used in oleoresinous varnishes only, some use is made of various unsaturated polyester resins for true 'twin pot' resins in which the resin and the initiator are sprayed simultaneously from twin-jet spray guns, to set on contact. It is possible, by this technique, to obtain films which are of the order of 0·2—0·5 mm in thickness. The reason why this can be done is that one does not need to use any solvents with this technique. Polyamide resins are sometimes used in conjunction with other resins, such as epoxy, to produce films that have good flexibility, adhesion and impact resistance.

9.6 Water paints

These can be subdivided into:

1 *Solution types*, in which the vehicle consists of a solution of the binder in water.

2 *Emulsion types*, in which the medium consists of an emulsion of the film former itself, which may be a drying oil, an oleoresin or a synthetic resin. In the latter case the non-polar material is insoluble in water but exists in the form of an oil-in-water emulsion, i.e. in the form of globules suspended in the water.

In the case of the solution types, the film produced has no resistance to water, whereas in the case of emulsion types, once the water has evaporated the film behaves to all intents and purposes like a normal paint film.

SOLUTION-TYPE WATER PAINTS

The main representatives of this category are whitewash and non-washable distempers. Whitewash is a suspension of powdered natural calcium carbonate in a dilute glue solution. In general, there is one part of glue to every 30 of calcium carbonate, and some preservative, such as phenol, cresylic acid or a chlorinated phenol is added, to stop the glue from putrefying. Whitewash is commonly used as a ceiling finish. It is unsuitable for walls because of its very poor film strength and its ready removal by water. In the case of non-washable distempers the binder consists of either glue or casein together with a small quantity of preservative. A very much higher binder/pigment ratio is used than with whitewash, and in consequence it becomes impossible to use powdered calcium carbonate alone as the main pigment as this would look dirty. The main white pigments used are lithopone (nZnS.mBaSO$_4$) together with calcium carbonate in the ratio of 1:1. Titanium dioxide is also used. As far as the coloured pigments employed in washable distempers are concerned, their main property must be that they are alkali-resistant. This means that neither Prussian blue nor chrome yellows can be employed.

EMULSION-TYPE WATER PAINTS

Emulsion paints are always oil-in-water emulsions which constitute suspension of oils, varnishes and synthetic resins in an aqueous medium. The continuous phase consists of an aqueous solution of surface-active agents and water-soluble colloids such as glue, casein, cellulose derivatives, etc. Preservatives to prevent putrefaction are always added. The emulsions can be thinned with water without either breaking or coagulating.

The simplest of all are oil-bound distempers, which are normally supplied in paste form and are thinned to requirements. The washability of such distempers is governed by the ratio of oil to glue. The lower the quantity of glue, the better the washability. Too much pigment also reduces the washability, while too little makes application difficult. Oil-bound distempers are widely used for interior work. They are generally not considered durable enough for external use.

Very durable water paints can be produced by using an aqueous emulsion of oil-modified alkyd resins. In such a case only a very small percentage of dispersing agent is added, and hence the washability is very good. Interior alkyd emulsion paints usually have lithopone as the main white pigment base, while exterior types tend to contain rutile titanium dioxide. About 0·1% cobalt drier is commonly employed calculated on emulsion solids.

For exterior use the pigment/binder ratio should at all times be kept below 0·5:1, while for interior use the ratio can be raised to 1·5:1. When a semi-gloss finish is required the pigment/binder ratio is usually about 0·2:1. As in the case of oil-bound distempers, interior finishes have lithopone as the main white pigment, while exterior ones contain rutile titanium dioxide.

Latex water paints consist of high-molecular-weight polymers that have been polymerized in the emulsified state. The most common of these are either polyvinyl acetate (PVAC) or PVAC–styrene–butadiene copolymer. The polymer is usually in the form of particles between 0·05 and 0·1 mm in diameter. If the particle size is larger than this, gloss is impaired, and there is a likelihood of sedimentation. In addition, a small particle size means that the emulsion has better water-resistance and gloss.

When PVAC emulsions dry, the polymer tends to form powdery particles unless properly plasticized with substances such as dibutyl phthalate or tricresyl phosphate. In addition, water paints often have thickeners added which are basically hydrophilic colloids, e.g. sodium carboxymethyl cellulose, various water soluble salts of polyacrylic acid, methacrylic acid and methyl cellulose. The thickeners increase the consistency of the emulsion paint and affect the flow.

Finally there are some polyacrylic and polystyrene base emulsion paints on the market, both of which produce durable water resistant films when dry.

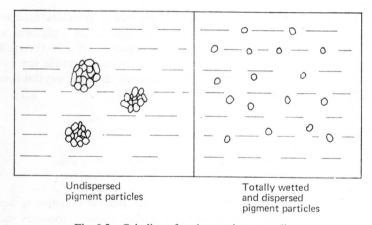

Undispersed
pigment particles

Totally wetted
and dispersed
pigment particles

Fig. 9.2   Grinding of a pigment into a medium

9.7 Pigments

Pigments are solid materials, which are suspended in the form of small discrete particles in the medium, and which serve to give the paint:

1 Colour.
2 Opacity.
3 Increased abrasion resistance.
4 Certain chemical properties.
5 Improved body.

Pigments can either be natural products which have been finely ground during manufacture, or they can be synthetic materials precipitated from solutions. They must at all times be individually suspended in the paint medium (see Fig. 9.2). This is done by employing a colloidal grinding technique. The pigment particles in the unwetted state tend to stick together and therefore a considerable amount of work has to be done on them in order to disperse them in the medium. The main types of grinders used in the paint industry to achieve this dispersion are the ball mill, the single roller mill and the triple roller mill.

Pigments can be divided most conveniently into the following types:

1 Whites.
2 Extenders.
3 Inorganic coloured pigments.
4 Organic and lake pigments.
5 Metal powders.

WHITES

Titanium dioxide

This is the commonest of all white pigments and is manufactured in two modifications: anatase and rutile. Its great advantages as a pigment are its very high refractive index, which is 2·52 for anatase and 2·76 for rutile, its chemical inertness and its lack of toxicity. The average particle size of TiO_2 pigments is about 0·25 μm. Rutile and anatase are different crystal forms of TiO_2 and have the following contrasting properties:

	Rutile	*Anatase*
Tinting strength	1,600	1,200
Density	4·2 kg/dm³	3·9 kg/dm³
Colour	Slightly cream	Pure white
Resistance to chalking	Very good	Poor

In consequence, titanium dioxide, with a high anatase content, is best for glossy paints which are used internally, while matt paints, especially those to be used externally, should have a much higher rutile content.

Titanium dioxide is used in every kind of paint with the exception of very dark coloured ones. It is particularly useful in paints that are used in circumstances where contact with food is likely.

Zinc oxide and zinc sulphide pigments

Zinc oxide has a refractive index of 2·08 and zinc sulphide one of 2·37. The density of both pigments is around 5·6 kg/dm³.

Zinc oxide is brilliant white, but has a much poorer covering power than titanium dioxide. In addition it is a relatively reactive pigment and when used with oleoresin finishes is often the cause for cracking on the film. Its main advantage is its property of being able to absorb ultraviolet light. This property is extremely advantageous for external paints in which delicate organic lake colours are used. The zinc oxide aids both colour retention and stability of the medium film. Zinc oxide on its own is non-toxic but it is frequently used in the leaded form, i.e. when it contains quantities of $PbO.2PbSO_4$. Such pigments must, of course, never be used where likelihood of contact with foodstuffs occurs, or where children may come into contact with the paint film. Zinc oxide is somewhat fungicidal and is thus useful in retarding the formation of mildew on paint surfaces. It also retards putrefaction in glues, adhesives and water paints. The material must never be used with acidic paint media as it will cause such media to 'thicken' prematurely.

Lithopone
This is made by the following process:

$$ZnSO_4 + BaS \rightarrow ZnS + BaSO_4$$

to produce a pigment which contains about 30% zinc sulphide. The material is then roasted and ground to produce a pigment with an average particle size of 0·5 μm.

Lithopone is a brilliant white pigment with excellent hiding power. Its main use is in water paints and washable distempers for internal uses, but it is also widely used in all types of paints, rubbers, plastics and flooring compositions. Lithopone should not be used for external purposes because it is liable to chalk, followed by weather erosion.

White lead
White lead has the general formula $Pb(OH)_2.nPbCO_3$ where n has a value between 1·5 and 3·6. It is mainly used today in primers or ground coats for wood and stucco. This pigment is blackened by exposure to hydrogen sulphide and its use is very much diminishing.

Antimony oxide
Antimony oxide (Sb_2O_5) has a refractive index of 2·0 and a density of 5·7 kg/dm³. Its main use in paints is to improve the brushing and levelling properties.

EXTENDERS

These are materials, which although they are usually white in colour, have a refractive index which is too low to contribute appreciably to the covering power of a paint. The reasons for the addition of extenders are the following:

1 They give body and improved brushing properties to a paint.

2 They increase the pigment content so as to produce a matt or semi-glossy surface suitable for undercoats.

3 Some of the extenders used help to keep the pigment in suspension.

The main extenders used are as follows.

Paris white or CaCO₃

This is a pigment which, when dispersed in oil or cellulose media, gives a dirty colour to the resulting paint. It can, however, be used in water paints and distempers as a vehicle. As calcium carbonate is alkaline, it cannot be used in conjunction with acid pigments.

Barytes and blanc fixe

These are both chemically the same, namely barium sulphate ($BaSO_4$). Barytes is the name given to the ground natural material, while Blanc Fixe is precipitated chemically. The latter has a finer crystalline shape. The disadvantage of both is the relatively high density of 4·6 kg/dm³, which tends to cause suspension difficulties.

Silica, slate powder and mica are all used as extenders in paints, where colour is not of primary importance. These materials are cheap and easy to incorporate.

Asbestine and talc are hydrated magnesium silicates with a theoretical formula of $H_2Mg_3(SiO_3)_4$. The density of the material is about 2·8 kg/dm³ and the refractive index 1·59. Both materials are widely used as suspending agents in paint and for anticorrosive surface coatings.

INORGANIC COLOURED PIGMENTS

There are two main types of inorganic blue pigments:

1 *Prussian blue*, made by reacting sodium ferrocyanide with iron salts. It is essentially $Fe_4[Fe(CN)_6]_3$ and is a strong pigment which is wholly unstable against alkalis. It must therefore never be used on setting concrete or plaster.

2 *Ultramarine blue* is a complex silicate which has a rather pinkish cast. It is stable towards alkalis but is attacked by acids. It is widely used in distempers and other water paints.

IRON OXIDE PIGMENTS

There is a large range of these pigments, which contain various oxides of iron together with additions of manganese dioxide. The colours are governed, not only by the actual chemical composition, but also by the state of subdivision, the crystalline forms, etc.

The main iron oxide pigments used in the paint industry are:

Red oxides of iron
Yellow ochres
Light brown siennas
Dark brown umbers

All iron oxide pigments are very durable and have good colour retention. They are chemically inert. They are able to absorb ultraviolet light and thus tend to protect external paint films.

Red lead

Red lead, which has the general formula $2PbO.PbO_2$, is used only in protective paints, being employed in primers for steel and iron work. Its protective action is largely due to its slight alkaline nature, and the fact that it forms weather-resistant lead soaps with the linseed oil paint base.

Chromate pigments

A series of chromate pigments is used, of which the most important are the following:

Primrose chrome : 50% lead chromate, 50% lead sulphate
Lemon chrome   : 75% lead chromate, 25% lead sulphate
Middle chrome   : 100% lead chromate
Orange chrome   : basic lead chromate
Scarlet chrome : lead chromate, lead sulphate and lead molybdate

All the lead chromates have the general disadvantages of lead pigments, which are poisonous nature and liability of blackening in the presence of hydrogen sulphide. Chrome pigments which do not suffer from these disadvantages are zinc chromate and barium chromate, also strong yellow in colour.

Black pigments

The great majority of black pigments used are carbon blacks, prepared by burning carbonaceous substances and allowing the flue gases to impinge upon cold surfaces. Several varieties of carbon black are on the market, their names indicating the carbonaceous substances used in their manufacture, e.g. lamp black, vegetable black, bone black, furnace black, etc.

A typical furnace black has the following properties:

Density	1.8 kg/dm^3
Average diameter of particles	30—60 nm
Volatile matter	0.5—1%
Surface area	40—90 m^2/g
pH	9.5

Carbon blacks have immense staining power but are hard to grind. Because they are so very light, they tend to float to the surface of paint films and in consequence paints which contain carbon black in addition to other pigments, tend to darken on standing. It is difficult to carry out accurate colour matching when using carbon black for this reason. If accurate tints are needed, organic lakes are used instead.

Graphite is also used to some extent as a black pigment, but only where its protective action against corrosion is to be utilized. It has a poor tinting strength and a low oil absorption figure.

M

ORGANIC PIGMENTS

Organic dyes all owe their colour to the fact that within the molecules there is a type of vibratory motion, which comes within the visible spectrum. Many of the common organic dyes dissolve readily in oil, resin or varnish and are used as such. In such circumstances no opacity is contributed, and the result is simply a coloured transparent film.

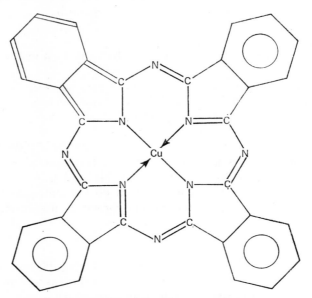

Fig. 9.3   Structure of copper phthalocyanine

Some directly produced dyes are quite insoluble in water or organic materials and can be used directly as pigments.

Typical of such pigments is the well-known blue copper phthalocyanine (see Fig. 9.3), made by reacting phthalic anhydride, urea and cupric chloride.

Lakes

Lakes are made by reacting typical organic dyes, which bear an active chemical group, with a carrier material, which may either be basic or acidic. There are many hundreds of organic lakes manufactured with widely varying colours and properties. Details should be sought from specialist literature.

METAL POWDERS

The main metal powders used in the paint industry are flake aluminium and zinc dust, of which the former is by far the most common. Aluminium used as a paint pigment may be either the leafing or the non-leafing kind and is available either as a powder, or as a paste in a suitable solvent such as white spirit.

The fact that the aluminium flakes form a continuous film on the surface of a paint medium provides good protection for the underlying surface and is particularly useful as a sealer. Because of the high reflectance for both visible light and ultraviolet light, the temperature of structures is reduced. Aluminium paint has excellent high-temperature resistance and is widely used for painting hot water pipes and radiators.

Zinc dust is often incorporated into exterior paints in conjunction with other pigments in order to produce special appearance effects. Its main use, however, is in protective paints for iron and steel structures where it acts as a type of sacrificial anode. Paints containing 80% Zn and 20% ZnO in oleoresin varnishes are used for painting galvanized surfaces, the insides of water tanks, etc.

9.8 Hot organic coatings

Surface coatings which are applied hot include the following:

1 Asphalts and related materials.
2 Synthetic hydrocarbon resins.
3 Cellulose derivatives.
4 Waxes of all types.

ASPHALTS AND RELATED MATERIALS

Natural asphalt, which is found in the USA, Trinidad and Venezuela, is a black material with a density between 0·9 and 1·0 kg/dm³ and a melting-point of between 76 and 86°C. It contains a mixture of wax esters, wax acids and resins with an average molecular weight of about 1,500.

Coal tar pitches have a higher density, which varies between 1·4 and 1·5 kg/dm³. The softening-point of coal tar pitches is of the same order as that of natural asphalts.

Both are widely used for the coating of submerged pipelines, for the production of roofing felt, and surface coating of concrete and other structures in permanent exposure to water. Asphalts have excellent resistance to most non-oxidizing acids but are attacked almost immediately by any non-polar solvent. Their resistance to alkalis is good, and to inorganic salts excellent.

SYNTHETIC HYDROCARBON RESINS

The main resins used for hot-surface coatings are as follows:

1 Coumarone – indene resins occur in light oil fractions of coal tar distillates. These are used for chipboard coatings, adhesives for aluminium sheeting to wood and similar purposes.

2 Terpene resins have molecular weights between 650 and 1,600.

3 Polyisobutylene and cyclized rubbers employed as hot coatings for materials of a pliable nature.

4 Polyethylene is applied hot and also has the property of good flexibility. The grades used for this purpose have molecular weights, varying from 1,000 to 9,000.

5 Polystyrene is blended with various waxes to produce hot applied coatings with high gloss, good elongation and excellent adhesion.

6 Chlorinated hydrocarbon resins, such as chlorinated biphenyl, terphenyl and naphthalene are blended with waxes to give flame-resistant coatings. The material has the trade name 'Hypalon' and is cured after application to surfaces.

CELLULOSE DERIVATIVES

The main cellulose derivatives used for hot coatings are cellulose acetate – butyrate resins plasticized with dibutyl and dioctyl sebacate. These materials have melting-points of between 120 and 140°C depending on the nature and quantity of plasticizer added. The material is widely used for hot dipping purposes, mainly for the surface protection of smaller parts. It is used in the building industry for coating door knobs and other fittings.

WAXES OF ALL TYPES

Waxes can be subdivided into vegetable, insect and mineral types. The composition of waxes varies very widely. Table 9.1 summarizes the main properties of commercial waxes.

TABLE 9.1

Wax	Melting-point, °C	Density, kg/dm³
Mineral wax, $C_{20}H_{42}$	37	0·777
Mineral wax, $C_{26}H_{54}$	56	0·778
Mineral wax, $C_{30}H_{62}$	66	0·780
Chlorinated paraffin wax, 70% Cl	86	1·64
Shellac wax	82	0·93
Bee's wax	50	0·97
Spermaceti wax	45	0·96
Carnauba wax	84	0·999
Candelilla wax	67—79	0·947
Sugar cane wax	80	0·983

Waxes are used both on their own and together with various resins as hot coatings to protect metals and timbers.

Literature Sources and Suggested Further Reading

1. BRAGOON, C. R., *Film Formation, Film Properties and Film Deterioration*, Interscience, New York (1958)
2. CHATFIELD, H. W., *Paint and Varnish Manufacture*, Newnes, London (1955)
3. CHATFIELD, H. W., *The Science of Surface Coatings*, Benn, London (1962)
4. EDWARDS, J. D., and WRAY, R. I., *Aluminium Paint and Powder*, Reinhold, New York (1955)
5. FISCHER, W. VON, *Paint and Varnish Technology*, Hafner, New York (1964)
6. GOSSETT, R. K., *Hot Organic Coatings*, Reinhold, New York (1959)
7. HARRISON, A. W., *The Manufacture of Lakes and Precipitated Pigments*, Leonard Hill, London (1957)
8. NYLEN, P., and SUNDERLAND, E., *Modern Surface Coatings*, Interscience, New York (1965)
9. *Paint Technology Manuals*, 6 volumes, Chapman and Hall, London (1960–1966)
10. PARKER, D. H., *Principles of Surface Coating Technology*, Interscience, New York (1965)
11. PAYNE, H. F., *Organic Coating Technology*, 2 volumes, Wiley, New York (1954)
12. REYNOLDS, W. W., *Physical Chemistry of Petroleum Solvents*, Reinhold, New York (1963)
13. ZAHN, E. A., *Paint Evaluation*, Research Press, New York (1955)

Chapter Ten **Plastics**

10.1 Introduction

One of the fastest growing fields of new materials in the building industry is that of plastics of all kinds. Basically plastics all consist of simple organic chemical molecules which have been induced by the action of light, heat or chemical initiation, or combinations of these, to form chains or three-dimensional structures (see Figs. 10.1—10.3). These are called *polymers*. The simple molecules from which these polymers are made are called *monomers*, and when different types of monomers are used for polymer formation the material produced is called a *copolymer*.

Polymers are of three main types:

1 Polymers in the form of long chains are popularly known as *thermoplastic* materials (see Fig. 10.4). The reason for this is that such materials can be softened by heat and reshaped if necessary. In addition such polymers can readily be dissolved in various solvents and mixed with *plasticizers*, materials whose function is to keep the plastic material soft and pliable.

2 Polymers in the form of two-dimensional or three-dimensional space lattices usually start as chains, but join together subsequently due to the formation of side-chain linkages (see Fig. 10.5). Such plastics are also called *thermosetting*, because after they have set, heating cannot melt them. Fully set plastics of this type cannot readily be dissolved in solvents, although they are usually soluble when in the partially set stage. Conversely, of course, the thermosetting plastics have the advantage of considerable solvent and other chemical resistance and are generally a good deal harder than the thermoplastic types of materials.

3 Polymers in the form of springy chains are commonly called elastomers. Originally the term rubber was given to the product obtained from the rubber tree, which produces a sap amounting to as much as 10% of the total weight of the plant. Today, however, the importance of natural

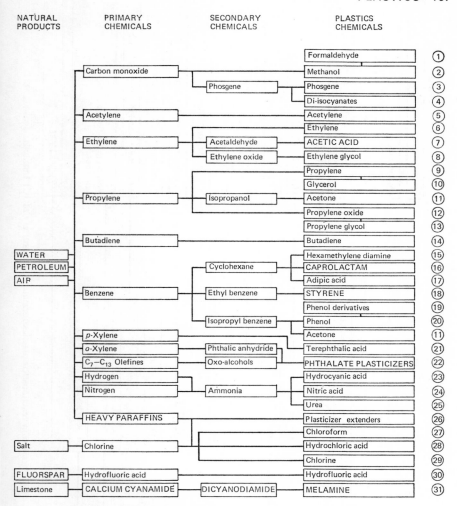

Fig. 10.1   Sources of plastics materials (By courtesy of ICI Limited)

rubber has declined considerably and most of the rubber of commerce is in fact a thermoplastic material, in which all or some of the molecules form zig-zag or spring-shaped chains. The distinction between rubbers and plastics is getting extremely obscure due to the large number of copolymers which are now being used between rubber-like structures and plain thermoplastic substances.

MOULDING OF PLASTICS

There are basically ten different methods of forming plastics. It should be realised that not all of these processes are suitable for each and every material,

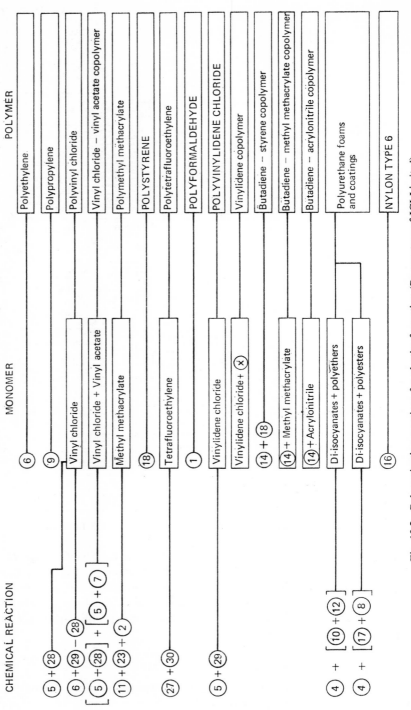

Fig. 10.2 Polymerization reactions for plastics formation (By courtesy of ICI Limited)

CHEMICAL   POLYMER
REACTION

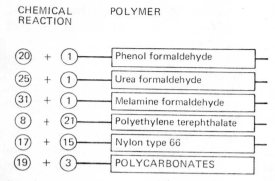

Fig. 10.3   Condensation reactions for plastics formation (By courtesy of ICI Limited)

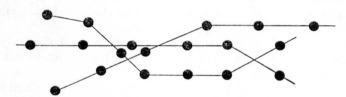

Fig. 10.4   Typical thermoplastic polymer

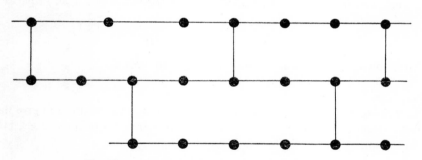

Fig. 10.5   Thermosetting polymer with cross-linkage

and that manufacturing techniques depend upon the chemical and physical nature of monomer and polymer.

The moulding methods are:

1 *Compression moulding* is nowadays used exclusively for thermosetting products in which the monomer is cured by heat and directly applied pressure.

2 *Injection moulding*, in which the plastic material is pushed into a fixed mould by a ram or screw, and cured inside this mould, to be ejected at the end of the curing period.

3 *Extrusion moulding* consists of forcing the softened material through an aperture to form tubes, rods and sheets.

4 *Casting* is a method of obtaining the material in liquid form by either melting it or dissolving it in solvents, and then shaping it without the application of pressure in open moulds.

5 *Calendering* is the production of sheets of material by rolling the plastic between multiple rollers.

6 *Rotational moulding* also employs a roller, but uses a knife to remove the thin film from its surface. This method is used for manufacturing thin plastic membranes.

7 *Spreading* is basically a hand process used for making large components in which the plastic mass is spread by means of a spatula on to a base mould. Initiation setting or infra-red radiation are the most common methods of further processing.

8 *Vacuum forming* is the shaping of sheets of material by means of a heated pressure mould or by vacuum. This process is one of the most important, as it can be used for shaping large sections.

9 *Blow moulding* is a process which simulates the manufacture of glass bottles or electric light bulbs, except that a highly viscous medium of plastic is used instead of molten glass.

10 *Laminating* is the process by which under the action of heat and pressure layers of plastic or layers of other materials such as wood, fabric, metal, etc., can be made into a composite whole, sandwiching a layer of plastic.

ADDITIVES TO PLASTICS

The following materials are commonly added to plastics.

Plasticizers
These are softening agents introduced into the plastic formulation to give the plastic material flexibility. Normal vegetable oils are used in some cases, but more usual are chemicals such as the following:

	ASTM abbreviation
Dibutyl phthalate	DBP
Dicapryl phthalate	DCP
Di-isodecyl adipate	DIDA
Di-isodecyl phthalate	DIDP
Di-iso-octyl adipate	DIODA
Di-iso-octyl phthalate	DIOP
Dioctyl adipate	DOA
Dioctyl phthalate	DOP
Dioctyl sebacate	DOS
Tricresyl phosphate	TCP
Trioctyl phosphate	TOF
Triphenyl phosphate	TPP
Dinonyl phthalate	DNP
Dioctyl azelate	DOZ
Di-n-octyl-n-decyl phthalate	DNODP

Reinforcing agents

Reinforcing agents are added to plastics in order to improve their toughness, especially from the point of view of increasing tensile strength. The most common type of reinforcing agent is glass fibre.

Extenders and fillers

Pure plastic materials are rather expensive and it is often found that the addition of a variety of inorganic and organic materials such as china clay, asbestine, carbon black, sawdust, etc., can be mixed with the plastic material without unduly affecting its properties. Often such materials serve a useful purpose as well by increasing the body of the plastic to aid in the moulding process, by giving the material the right degree of opacity and also by increasing the fire-resistance of the material.

Colouring matter

This is usually in the form of pigments and is often added to the plastic monomers.

THE CHEMICAL CHARACTERISTICS OF PLASTICS

The number of molecules of monomer which form a polymer varies very widely even within one given plastic mass. Often the physical and chemical properties of the final product are governed largely by the extent to which this polymerization or *curing* process has taken place.

Typical degrees of polymerization and molecular weights are given in Table 10.1.

TABLE 10.1

Material	Number of monomers per polymerized molecule	Range of molecular weights
Polystyrene	1,400—3,500	90,000—200,000
PVC	4,000	250,000
Natural rubber	2,000—6,000	140,000—420,000
Nylon 66	90— 150	18,000— 30,000

METHODS OF POLYMERIZATION

The tendency of plastic monomers to form polymers depends markedly upon their chemical composition. The most important feature of all is the possession of one or more double bonds in the compound, as for example:

Ethylene	$H_2C:CH_2$
Vinyl chloride	$CH_2:CHCl$
Vinylidine chloride	$CH_2:CCl_2$

Acrylonitrile	$CH_2:CHCN$
Vinyl acetate	$CH_2:CHOCOCH_3$
Methyl acrylate	$CH_2:CHCOOCH_3$
Methyl methacrylate	$CH_2:C(CH_3)COOCH_3$
Isobutylene	$CH_2:C(CH_3)_2$
Styrene	$CH_2:CH.C_6H_5$
Tetrafluoroethylene	$F_2C:CF_2$

The ease of polymerization depends markedly not only upon the number of double bonds in the molecule but also upon the type of substituents. For example, vinylidine chloride polymerizes much more rapidly than vinyl chloride and methyl acrylate polymerizes more rapidly than methyl methacrylate.

Apart from polymerization which takes place via double bonds, which is by far the most common, polymerization may also take place via so-called organic condensation reactions. This takes place when two substances have both disposible hydrogen and hydroxyl groups attached. Under the action of heat or initiation these react together to form water, leaving the remaining groups joined. These reactions apply in many thermosetting compounds such as polyesters, phenolics and urea formaldehyde types of plastics.

INITIATION OF REACTIONS

To enable a polymer reaction to take place, the energy content of the final polymer must be less than the energy content of the monomers taking part in the reaction. This means that in most cases the polymerization process goes forward with the emission of heat energy. However, energy is often needed to initiate the reaction and is usually supplied in the form of heat.

Thermal or photochemical initiation creates active molecules either by impact or directly. These active molecules react with others to form chains.

Polymerization reactions can also be initiated chemically.

Free-radical initiation
This produces an entity, where one valency is momentarily incomplete and such a material is by its nature chemically extremely active. The most common free radical initiator is benzoyl peroxide (Fig. 10.6). This splits up at a temperature above 100°C to form momentarily the free radical $C_6H_5COO\cdot$, which then splits up further into CO_2 and $C_6H_5\cdot$, a very powerful free radical able to initiate the formation of such plastic materials as polystyrene. Other initiators which work in a similar way are organic hydroperoxides such as t-butyl hydroperoxide $(CH_3)_3C{-}O{-}O{-}H$. (The $\equiv C{-}O{-}O{-}H$ group is not to be confused with the $-COOH$ group which is the normal acid grouping.) These are used in the initiation of polymerization of vinyl compounds.

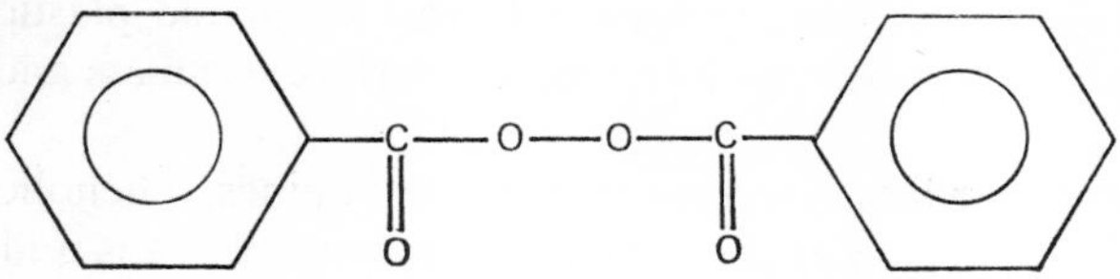

Fig. 10.6 Structure of benzoyl peroxide

Ionic initiation
This type of initiation makes use of materials such as stannic chloride, titanium trichloride, aluminium chloride and other materials which serve to initiate polymerization by polarizing the double bond of the monomer, i.e. by tending to acquire a pair of electrons from this double bond. The polymers which are most readily affected by ionic initiators are isobutylene, styrene compounds and vinyl ethers.

Initiation by electron donation
Polymerization is also readily initiated by certain *electron donating* substances such as alkali metal amines, e.g. $NaNH_2$, and the Grignard reagents which are ethereal solutions of magnesium alkyl halide compounds. The anionic fragment of the molecule repels two electrons of the double bond and thus induces a negative charge at the active end of the growing chain.

Redox initiation systems
Finally one may use a *reducing/oxidizing* (Redox) system to initiate polymerization reactions. As the oxidizing agent is being reduced, a free radical is formed which then serves to initiate the polymerization reaction. Typical examples are:

1 Ferrous sulphate and hydrogen peroxide form free OH radicals:

$$Fe^{2+} + H_2O_3 \rightarrow Fe^{3+} + OH^- + OH\cdot$$

2 Sodium persulphate and sodium thiosulphate form free $S_2O_3^-$ and SO_4^- radicals:

$$S_2O_8^{2-} + S_2O_3^{2-} \rightarrow SO_4^{2-} + SO_4^- \cdot + S_2O_3^-.$$

Such types of initiators are not as effective as peroxide initiators and setting takes place only slowly. Redox initiators are widely used for setting plastic glues.

10.2 Physical characteristics of plastics

HARDNESS OF PLASTIC MATERIALS

When plastics are used in the building industry, surface hardness is often one of the main criteria, owing to the need to avoid wear.

From the hardness point of view one can subdivide plastics into three groups: high surface hardness, intermediate surface hardness and low surface hardness.

High surface hardness is possessed by melamines, phenolics and urea-formaldehyde resins, and of these melamine formaldehyde is undoubtedly the hardest, hence its use for purposes where good abrasion resistance is of vital importance.

Polyester resins, epoxy resins, acrylic resins, glass-fibre-filled nylons, poly-carbonates and the general-purpose polystyrene materials have intermediate surface hardnesses. On the Rockwell M scale, melamines are given a figure of between 110 and 125, while general-purpose polystyrene has a figure of 70—80.

Low surface hardness is possessed by cellulose acetate, Acetal, ABS resin, nylon-6 and nylon-6.6, cellulose acetate butyrate, and polypropylene.

Plastics as a whole have a lower tensile strength than metals, but due to their general low specific gravities, their strength to weight ratios are often a good deal higher. Table 10.2 gives the tensile strengths and the comparative strength to weight ratios of several materials. It can be seen from this table that glass-fibre-reinforced plastics have particularly good strength to weight ratios.

TABLE 10.2

Material	Tensile strength, MN/m^2	Density, kg/dm^3	Strength/weight ratio, $10^3 m^2/sec^2$
Alloy steel	1,000	7·8	128
Titanium	800	4·5	178
Copper	350	8·9	39
Lead	15	11·3	1·3
Phenolic resins	50	1·5	33
Melamines	70	1·5	47
ABS	40	1·05	38
Acrylic resins	60	1·2	50
Polystyrene	50	1·1	45
Polyesters	45	1·05	43
Glass fibre	1,400	3·0	470
Filament-wound epoxy	450	1·3	350
Glass-fibre-reinforced epoxy	200	1·2	165
Glass-fibre-reinforced phenolics	240	1·8	135
Glass-fibre-reinforced polyesters	160	1·3	125
Glass-fibre-reinforced nylon	110	1·1	100

10.3 Rigid plastics used in the building industry

A very large number of different types of plastics have been developed during the last few years. Not all are used in the building industry, but up to now a use has been found for the following types of rigid plastics:

1 Phenol formaldehyde and related materials (PF).
2 Melamine formaldehyde (MF).
3 Rigid polyvinyl chloride (PVC).

4 Cellulose derivatives. These incorporate the following main types: carboxymethyl cellulose (CMC); cellulose acetate (CA); cellulose acetate – butyrate (CAB); cellulose acetate – propionate (CAP).
5 Acrylonitrile – butadiene – styrene copolymers (ABS).
6 Polymethyl methacrylate (PMMA).
7 Polystyrene (PS).
8 Polyester resins (numerous types).
9 Epoxy resins (numerous types).
10 Polypropylene (PP).
11 Nylon (6.6, 6.10, 11, 12, etc.).

PHENOLICS

Phenol formaldehyde plastics are formed by condensation reactions (elimination of water) between phenol C_6H_5OH and formaldehyde HCHO (see Figs. 10.7—10.9). Chains are formed at first and the material may then be readily dissolved in solvents. At this stage the substance is usually mixed with such fillers as chopped fabric, cord and glass fibre. Phenolics also incorporate such related materials as resorcinol formaldehyde and phenol furfural resins. The main methods of moulding used are compression and injection moulding, although extrusion moulding is used for the manufacture of rods and tubes. Moulding temperatures are between 140 and 180°C, but higher temperatures can be used for high-speed moulding processes. The mould shrinkage is lowest with cellulose-filled phenolics, when a figure of around 0·6% is obtained. With synthetic fibre-filled materials markedly higher mould shrinkages, of the order of 1·5%, are sometimes found.

The mechanical properties of phenol formaldehyde vary very considerably with the addition of different fillers, extenders, reinforcing agents, etc., as well as with the curing temperature employed.

The following are the data which apply to common general purpose phenol formaldehyde plastics:

Density	1·3 to 1·6 kg/dm³
Thermal stability	up to 150°C
Ultimate tensile strength	35 to 60 MN/m²
Water absorption in 24 hr	0·3—0·8%
Coefficient of linear thermal expansion	3×10^{-5} degC⁻¹

The inflammability of the material varies according to the filler used. The grades filled with asbestos and glass fibre are non-flammable, while the wood-flour- and cotton-filled materials have slight inflammability. The electrical volume resistivity also varies with the filler used and ranges from 10^9 to 10^{14} Ω.cm.

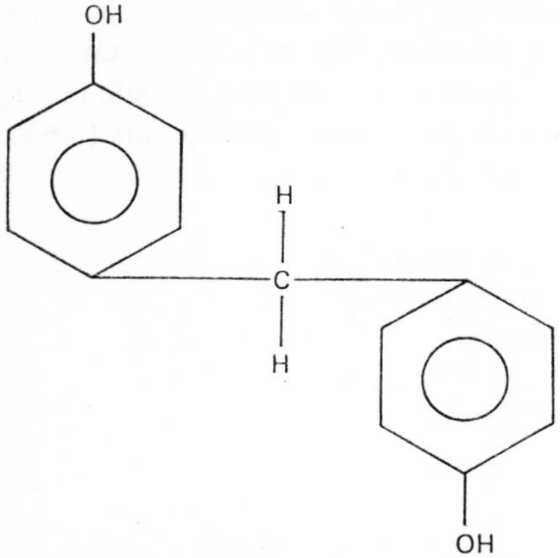

Fig. 10.7   First stage in the production of phenolic resins

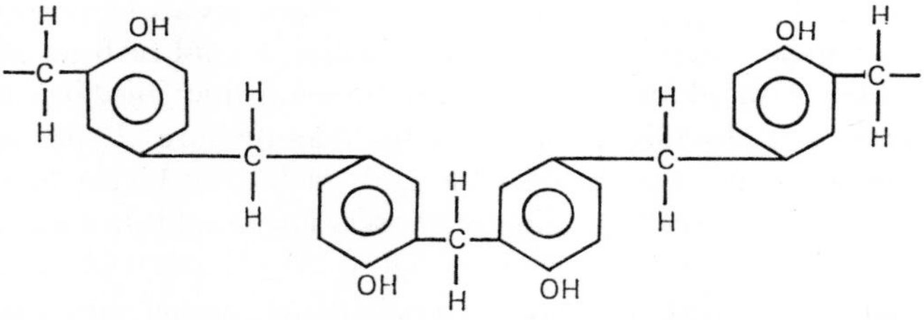

Fig. 10.8   Second stage in the production of phenolic resins

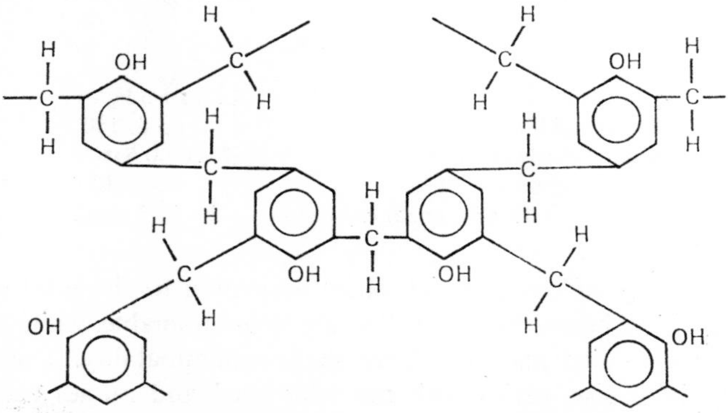

Fig. 10.9   Third stage in the production of phenolic resins

Chemical resistance

Phenol formaldehyde has, in general, a poorer chemical resistance than phenol furfural, which resembles it (Fig. 10.10). Where good chemical resistance is needed, the latter should be used instead.

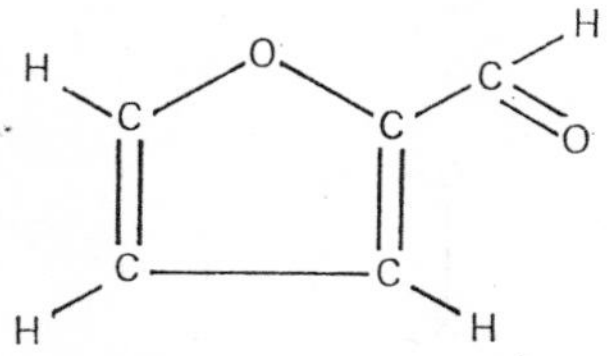

Fig. 10.10 Structure of furfural

Phenol formaldehyde has the following chemical resistances:

Dilute mineral acids	Fair
Concentrated mineral acids	Poor
Oxidizing acids	Very poor
Alkalis	Poor to very poor
Alcohols	Good
Aromatic hydrocarbons	Good
Chlorinated hydrocarbons	Good
Ketones	Fair
Soaps and detergents	Fairly good
Oils, waxes and greases	Good

The general weathering properties of phenol formaldehyde are excellent, but the surface hardness is somewhat affected by hot and damp conditions. It tends to crack when continuously exposed to steam at temperatures above 100°C.

Uses

Phenolic types of plastics are very widely used for making all types of electrical fittings and equipment, furnishing components and small fittings. They are also used for the manufacture of laminates. The most widely used of these is based upon phenolic impregnated Kraft paper, which has a decorative surface impregnated with melamine. Phenolic materials are also used for the manufacture of prefabricated building components, while resorcinol formaldehyde is most widely used as a glue.

N

MELAMINE FORMALDEHYDE

Melamine is a compound which behaves somewhat like urea, and when reacted with formaldehyde under neutral or slightly alkaline conditions, gives methylol melamines which are the monomers for the formation of the normal melamine moulding powders (see Fig. 10.11). The pure resins are quite

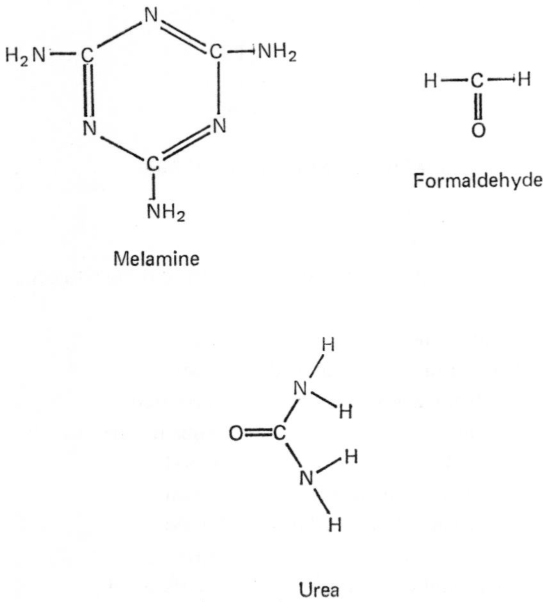

Melamine

Formaldehyde

Urea

Fig. 10.11   Structures of melamine, formaldehyde and urea

colourless and for this reason melamine can be formed into light and pastel shades if required. Melamines are thermosetting resins only. Melamine casting powders are usually supplied in four modifications, namely wood filled, cellulose filled, mineral filled and glass-fibre filled. The densities of the first two types when cured are between 1·5 and 1·55 kg/dm^3, while the other two have densities between 1·8 and 2·1 kg/dm^3, respectively.

Melamines are usually processed at temperatures between 140 and 170°C and the usual method of processing is compression moulding only. For curing of melamines, pressures of between 200 and 600 kg/cm^2 are used and very short curing periods are generally only needed. Care must be taken not to over-cure as this causes embrittlement and scorching. The maximum working temperatures of the types filled with organic fillers is around 100°C, while the types that are filled with mineral materials can be used continuously at temperatures up to 150°C.

Melamines are self-extinguishing from the point of view of flame-resistance and have a thermal conductivity varying between 0·08 and 0·15 W/m.degC, again depending upon the nature of filler used.

Chemical resistance
This varies with the type of filler used. Table 10.3 gives the main chemical resistances for (a) organic-filled melamines and (b) inorganic-filled melamines.

TABLE 10.3

	Chemical resistance of melamines	
Chemical	Organic filled	Inorganic filled
Mineral acids (dilute)	Good	Variable
Mineral acids (concentrated)	Poor	Poor
Alkalis	Good	Fair
Alcohol	Good	Fair
Ketones, aromatic hydrocarbons and chlorohydrocarbons	Good	Fair
Oils and greases	Good	Fair
Detergents and soaps	Excellent	Good

In general, the glass-fibre-reinforced melamines have chemical resistances which are almost as good as those of the organic-filled melamines.

Uses
Melamine formaldehydes are used for most purposes for which urea-formaldehyde plastics are used, but tend to be rather more expensive than the latter. They have good surface hardness, scratch resistance, colour stability and electrical insulation properties.

They are used for decorative laminates, door and cupboard fittings, housings for appliances and sanitary fittings. Melamine formaldehyde is also used widely as a coating resin, but must be plasticized, preferably with an oil-modified alkyl resin as it otherwise forms a brittle film.

RIGID POLYVINYL CHLORIDE (PVC)
Polyvinyl chloride is made in two modifications, the rigid type and the flexible type. The latter will be dealt with in Chapter 11. Polyvinyl chloride and polyvinylidine chloride are formed as follows:

$$n\text{CH}_2\text{:CHCl} \rightarrow (-\text{CH}_2-\text{CHCl}-)_n$$
vinyl chloride   polyvinyl chloride

$$n\text{CH}_2\text{:CCl}_2 \rightarrow (-\text{CH}_2-\text{CCl}_2-)_n$$
vinylidene chloride   polyvinylidene chloride
(vinyl dichloride)

Vinyl chloride monomer, which is a liquid boiling at $-14°C$, is unstable in the presence of oxygen. It does not polymerize thermally and only very slowly under photochemical action. Polymerization is initiated readily at temperatures between 30 and 80°C, and since the polymer is insoluble in the monomer liquid, it precipitates as soon as it is formed. The rate of polymerization is always slow at first, but accelerates markedly, after a time reaching a maximum after 50% conversion.

PVC is a thermoplastic material and can be processed by injection moulding, extrusion, blowing, vacuum forming, rolling and also by compression moulding. Shapings can be fabricated from sheet PVC, which is easily welded and glued. Rigid PVC is a hard and horny material which is insoluble in a large number of solvents and not easily softened. It differs from the flexible type of PVC in that it has little or no plasticizer added.

When PVC is in the molten state it still possesses a considerable viscosity. This viscosity is non-Newtonian (it varies with the rate of shear) and at 180°C it varies from 100—900 kg.sec/m depending upon the rate of shear. Processing temperatures range between 150 and 185°C and mould shrinkage amounts to about 0·7—1·5%. Very thin sections, with thicknesses below 1·5 mm, are difficult to make in PVC because of the high melt viscosity of the monomer. Rigid PVC is made in three general grades:

1 Normal impact with a density of 1·39 kg/dm³.
2 High impact with a density of 1·34 kg/dm³.
3 High-temperature grade with a density of 1·54 kg/dm³.

Various additives are usually incorporated in PVC, and the properties of the material are altered considerably when this is done.

When rigid PVC is tested according to the BS 2782:1956 softening-point test, the first two grades are found to soften appreciably at 80°C, while the high-temperature grade can withstand a temperature of up to 120°C. Heat distortion, however, starts with all grades at between 70 and 90°C. PVC is self-extinguishing from the fire-resistance point of view.

Chemical properties
Rigid PVC is poor from the water absorption point of view. When immersed in water at atmospheric temperature for 24 hr the two first grades absorb as much as 15% by weight. The high-temperature type of PVC is fairly water-resistant, the increase in weight after 28 days of water immersion being only of the order of 0·5%. At 100°C, however, the corresponding weight gain amounts to more than 14%.

PVC also has a fairly high gas permeability, a typical figure for carbon dioxide being $2·4 \times 10^{-13}$ m⁴/kN.sec.

All forms of PVC have excellent resistance to both dilute and concentrated mineral acids and to alkalis.

The chemical resistance to alcohols is also excellent for all except the high impact grade.

The chemical resistance to ketones and aromatic hydrocarbons is poor for the high impact grade but fair to good for the other two grades. Similarly, all grades have excellent resistance to soaps and detergents and good resistance to oils and greases. The weatherability of PVC is excellent.

Uses

PVC is used in the building industry for the construction of building panels, wall tiles, external sheeting of all kinds such as sidings, roof components, etc. It is commonly used for window components and such fittings as ducts, vents, rainwater gutters and down pipes. Polyvinyl dichloride, which is also used in rigid modifications, has a working temperature some 25—35°C higher than normal PVC, and in addition it possesses a higher chemical resistance and greater strength. The material can be used for hot and cold water piping.

CELLULOSE DERIVATIVES

Cellulose is a complex carbohydrate, with the formula $[C_6H_7O_2(OH)_3]_n$ obtained from cotton linters, wood pulp tissue or cotton waste. Its main derivatives used in the plastics industry are:

1 Cellulose nitrate.
2 Cellulose acetate.
3 Cellulose propionate.
4 Cellulose acetate – butyrate.
5 Methyl cellulose.
6 Ethyl cellulose.

Of these, *cellulose nitrate* is not widely used in the building industry, due to the fact that the material is very inflammable.

Cellulose acetate

Cellulose acetate is made by the reaction of cellulose with acetic anhydride. For plastic moulding purposes the acetyl content is between 36·5 and 38·5%, which means that between 2·2 and 2·3 ester groups per molecule are present. Cellulose acetate is fairly non-flammable and is suitable for injection moulding. It can be moulded without a plasticizer but then possesses a very high working temperature. Most of the useful cellulose plastics contain between 60 and 85% cellulose acetate.

Cellulose acetate does not resist water too well and for this reason it should not be employed where good weathering properties are needed. Resistance to dilute acids, petrol and oils is good.

Cellulose propionate

This is somewhat similar to cellulose acetate, except that it requires less plasticizer than the acetate for the same flexural strength and has a higher impact strength than the latter. Its use in the building industry is increasing for many internal applications, such as light fixtures, light shades, housings, toilet seats, etc.

Cellulose acetate–butyrate

Cellulose acetate–butyrate is a mixed ester and combines the thermal stability of cellulose acetate with the improved water resistance and the higher thermal stability of the butyrate. Acetate–butyrate is found to be of particular value for piping, especially for the transportation of petroleum products. It is also used for the construction of many internal fittings such as telephone housings.

Methyl and ethyl cellulose

These come under the category of cellulose ethers. Of the two, ethyl cellulose is the more common. It is made by reacting ethyl chloride and caustic soda with cellulose. Its main feature is that it has a density of only $1·14$ kg/dm^3 and the most common grades have an ethoxy content of between 44 and 50%. Ethyl cellulose is stable down to temperatures of $-40°C$ and can be compounded with a large variety of plasticizers, resins, etc. Its resistance to chemicals is as follows:

Dilute acids	Fair
Concentrated acids	Poor
Alkalis	Good
Petrol and oil	Good
Aromatic solvents, esters, alcohols, etc.	Poor

Ethyl cellulose is mainly used for such purposes as artificial leather, flexible rubber-like tubing and similar.

ACRYLONITRILE – BUTADIENE – STYRENE COPOLYMERS (ABS)

ABS, as it is commonly called (see Fig. 10.12), is available in five grades, namely:

> Extra high impact.
> High impact.
> Medium impact.
> High heat.
> Metal plating.

The plastic material comes in the form of pellets, which can be injection moulded, extruded or blow moulded. The tensile strength of the material varies with grades, the extra-high-impact grades having the lowest UTS, i.e. 34 MN/m^2 at 23°C, while the high heat grades have a UTS of 50 MN/m^2 at this temperature. The density of ABS is very low, about $1·04$ kg/dm^3. It is distorted at a temperature of just 100°C when loaded and burns slowly when ignited, at a rate of 4 cm/min without dripping. Water absorption at 20°C is $1·05\%$ and gas permeability for air at 25°C is $1·3 \times 10^{-13}$ m^4/kN.sec.

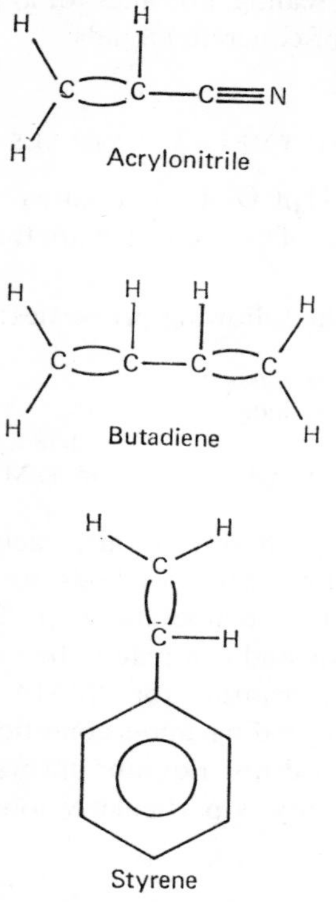

Fig. 10.12   Structures of acrylonitrile, butadiene and styrene

ABS when pigmented black, weathers extremely well and about 80% of its impact strength is retained after 3 years' exposure to strong sunlight. Light colours do not weather quite so well. Even so, its tensile strength is virtually unaffected by 3 years' exposure, although the impact strength may be cut to half.

Chemical resistance

Dilute mineral acids	Good
Concentrated hydrochloric and phosphoric acids	Good
Concentrated nitric and sulphuric acids	Poor
Alkalis	Very good
Detergents	Very good
Oils and greases	Good
Solvents	Poor

From the point of view of its use in the building industry, ABS has the advantages of being tough and rigid with fairly good heat stability and relatively low cost. It is used for drain, waste and vent piping, sliding doors and window tracks, weather sealing, housings for appliances and also, increasingly, for the manufacture of concrete moulds.

POLYMETHACRYLATES

Methacrylic acid, $CH_2:C(CH_3)COOH$, is used as the parent compound for the manufacture of a number of esters, which are the monomers of the methacrylate types of plastics.

Polymethacrylates have the following properties:

Light transmission	92%
Refractive index	1·49%
Density	1·18 kg/dm³
Tensile strength	up to 60 MN/m²

Methacrylic sheet may be shaped by air, vacuum forming, mechanical pressures or by a combination of these methods, using a temperature of about 160°C. For injection moulding, temperatures of 230°C are commonly used. The material is thermoplastic and can readily be welded or glued.

The maximum service temperature for PMMA (polymethylmethacrylate) is between 65 and 95°C, depending upon conditions. PMMA is difficult to ignite but burns slowly with drips. Equilibrium water absorption at 20°C is 2·1% and the gas permeability is particularly low, having a value of only 10^{-15} m⁴/kN.sec.

Chemical resistance

Dilute mineral acids and alkalies	Excellent
Concentrated acids	Fair
Detergents, soaps, greases, oils, etc.	Excellent
Nearly all solvents	Very poor

Methacrylates have a good resistance to weathering, being hardly affected at all by even tropical sunlight conditions. They are used in the building industry mainly for signs, light fittings, sanitary ware such as baths, wash basins, etc., canopies, casings and numerous other indoor and outdoor features. PMMA is also widely used instead of glass, where glass cannot readily be employed because of its brittleness.

POLYSTYRENE

Styrene is a very simple aromatic compound with an unsaturated side chain and the formula $C_6H_5.CH:CH_2$ (Fig. 10.13).

It readily polymerizes into a thermoplastic chain under the action of heat or an initiator. Commercial polystyrene transmits some 90% of white light and has a softening-temperature of just above 100°C. It is distorted by heat above 75°C and has a density of only 1·05 kg/dm³. Its tensile strength is of the order of 40—55 MN/m².

Polystyrene is normally injection moulded at temperatures between 160 and 280°C, using a minimum moulding pressure of 1·4 MN/m². Polystyrene is extremely rigid but relatively brittle. It suffers long elongation at break and

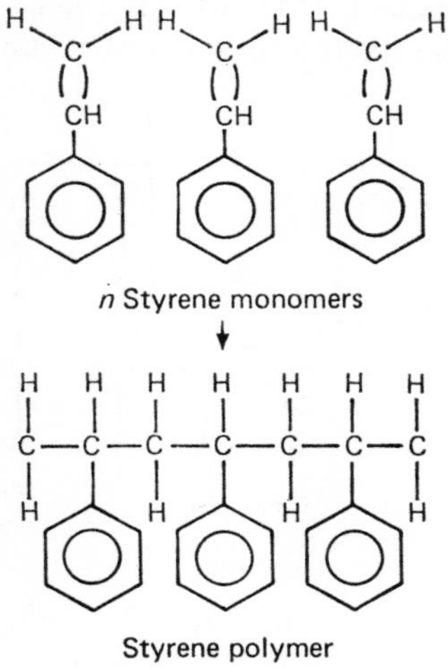

n Styrene monomers

Styrene polymer

Fig. 10.13   Styrene polymer production

parts must therefore be designed to work at low strains. Polystyrene is readily crazed by most solvents and oils, which very much increase the liability of the material to cracking. Polystyrene has good creep resistance but extremely poor dynamic fatigue resistance. Its electrical properties are excellent and are virtually unaffected by humidity. The material burns at a rate of 25 mm/min without the formation of drips and forming dense smoke.

Chemical resistance

Concentrated and dilute mineral acids	Very good
Oxidizing acids	Fair
Alcohol	Good
Organic solvents of all kinds, oils and greases	Poor
Soaps and detergents	Good

Uses

Apart from the use of polystyrene as a component of elastomers and foamed products, which will be dealt with in Chapter 12, the main use in the building industry is for wall tiles, light fixtures, knobs and drawer pulls, and similar fittings. The main advantages of polystyrene are its low cost and good transparency, while its main drawback from the point of view of use in the building industry is its tendency to craze and crack, coupled with poor impact resistance and poor weathering properties.

The unsaturated polyester resins polymerize by cross-linking among the individual linear polymer chains (Fig. 10.14). The advantage of polyesters is the fact that, unlike other thermosetting resins, no by-product is formed

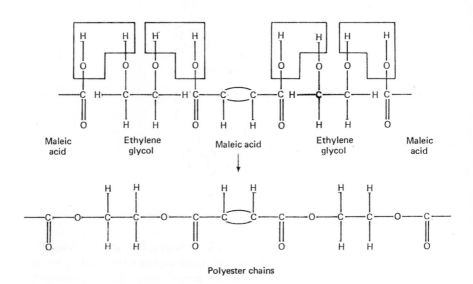

Fig. 10.14   Production of polyester chains

during the curing reaction and thus the resins can be moulded, cast and laminated at low pressures and temperatures. This not only simplifies the design of the moulds necessary, but also permits large articles to be fabricated, which could not normally be produced because of the high cost of making pressure moulds.

Polyester resins used for the production of rigid sections consist of:

1 Unsaturated dibasic acids which are condensed with saturated glycols, or vice-versa.

2 Unsaturated vinyl-type monomers which bridge across these strings.

Although a very large number of different combinations could theoretically be used, the actual number employed is limited.

A common polyester resin uses ethylene glycol – maleic acid chains and has styrene as a cross-linkage agent (Fig. 10.15). Such resins, however, tend to be somewhat brittle and for this reason part of the maleic acid is often replaced by such acids as phthalic, adipic or sebacic.

Another group of polyesters uses diallylphthalates. Polyesters are nearly always polymerized by chemical initiation, peroxide initiators being the most common.

If it is necessary to polymerize the polyester at temperatures below 60°C, acetyl benzoyl peroxide is usually employed as an initiator. At a temperature between 60 and 100°C lauroyl peroxide or benzoyl peroxide is used, while at higher temperatures materials such as dibenzaldiperoxide are used.

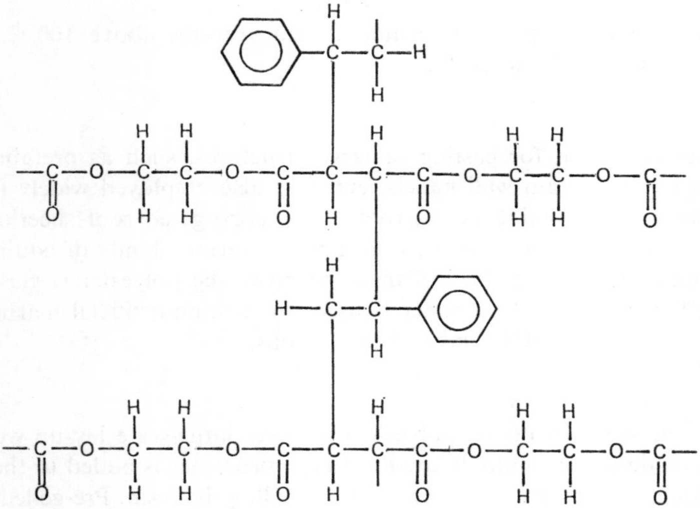

Fig. 10.15 Cross-linkage of polyester with styrene

Unsaturated polyester resins are commonly coordinated with many kinds of inert mineral fillers and with reinforcing fillers such as glass, paper and cloth. Of particular importance are the woven or free glass-fibre filaments which are added to produce plastic sections of very high strength and durability.

Physical and chemical properties
Owing to the large variety of commercial polyester resins which are used, the physical and chemical properties given here of necessity represent purely average values. Densities vary between 1·03 and 1·12 kg/dm³ and the materials have good stress crack resistance. Mould shrinkages of most of the resins are of the order of 8—10% but casting polyester resins have shrinkages of only 5%. The casting resins, however, cannot be used above 55°C as they tend to soften. Normal polyester resins have service temperatures between 95 and 220°C, depending upon type. The unreinforced resin has a thermal conduc-

tivity of 0·22 W/m.degC and a refractive index of 1·56. The equilibrium water absorption figure lies at 3% at 20°C.

Chemical resistance

Dilute mineral acids	Good
Concentrated acids	Fair
Alkalis	Poor except for chemically resistant grades
Detergents, soaps, greases and oils	Good
Ketones and chlorinated hydrocarbons	Poor
Alcohols and aromatic hydrocarbons	Fair

Polyesters, when exposed to water at temperatures above 100°C, show considerable tendencies to craze.

Uses

The material is used for casting of large structures, such as prefabricated bathroom pods, curtain wall panels, etc. It is also employed widely for the manufacture of swimming pools, roof lights, corrugated roof sheeting, flat translucent and decorative sheet, housings for various kinds of equipment, tanks, pipes and ducting. For all these purposes the polyester is glass-fibre reinforced. Unreinforced polyester is used for making artificial marble surfaces, for flooring, mortars and embedding jobs.

Methods of casting polyesters

Much of the work involving polyesters involves large-scale lay-up work. A pre-gel technique is commonly used in that a promoter is added to the resin which reduces the time usually required for gelling the resin. Pre-gelled structures have sufficient rigidity to be handled and a final cure can usually be carried out by heat treatment. In some cases no such heat treatment is necessary. Polyesters can be coloured with light-fast and durable pigments. When glass fibre is added as a reinforcing agent, products are obtained which have very considerable strength but weigh only 35% as much as similar steel objects and only 70% as much as similar aluminium objects. The polyester castings have very considerable impact resistance and will not normally crack. If, however, they should somehow be damaged they can be readily repaired by using the same unsaturated polyester resin mixed with initiator, as used for the original structure.

EPOXY RESINS

Epoxy resins are ether polymers that contain phenylene groups, as well as alcoholic groups. The intermediates are usually made by reacting such substances as a diphenol with epichlorohydrin (see Figs. 10.16 and 10.17). These are either viscous liquids or low-melting-point solids. The materials are then

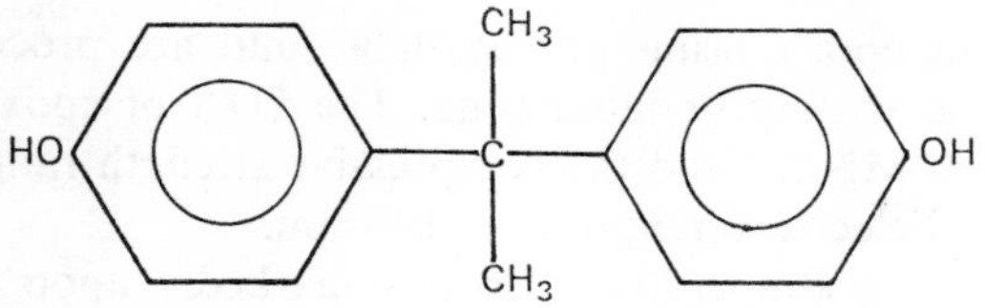

CH$_3$

HO〈◯〉—C—〈◯〉OH

CH$_3$

Bisphenol A

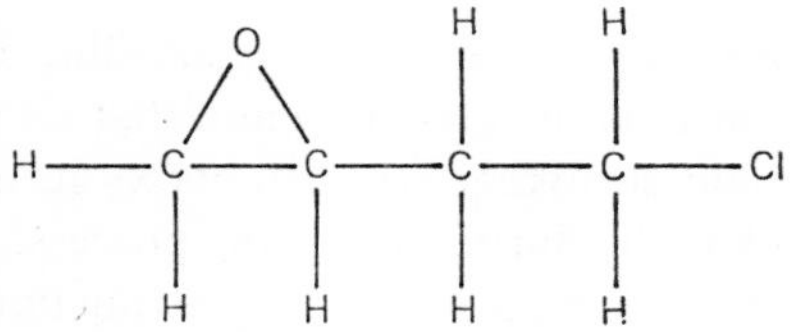

O
/ \
H——C——C——C——C——Cl
| | | |
H   H   H   H

with H H on the upper C—C positions

Epichlorohydrin

Fig. 10.16 Structures of bisphenol A and epichlorohydrin

converted into thermosetting resins by cross-linking these chains with various organic acids, amines and similar substances. Curing can take place either at room temperature, a process which takes place readily when 5—10% of amines are added to the epoxy monomers, or at higher temperatures. Epoxy resins are widely used as adhesives to provide extremely strong bonding between all kinds of materials. When used as an adhesive, the epoxy resole is kept separate from the curing agents until needed. After mixing the reaction between the two agents to form a rigid thermoset plastic structure proceeds slowly in the cold, and much more rapidly at elevated temperatures.

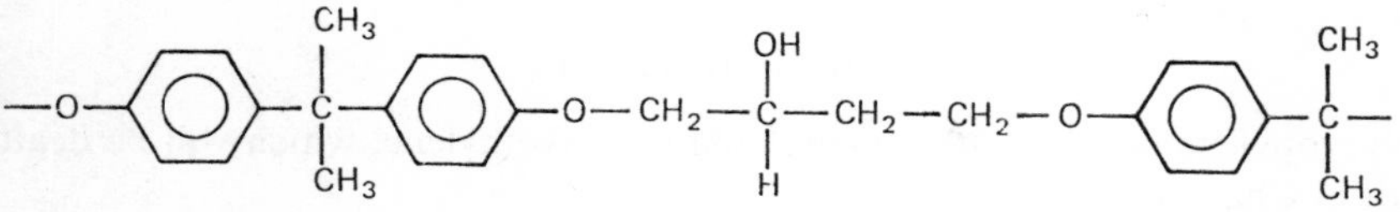

CH$_3$ OH   CH$_3$
| | |
—O—〈◯〉—C—〈◯〉—O—CH$_2$—C—CH$_2$—CH$_2$—O—〈◯〉—C—
| | |
CH$_3$ H   CH$_3$

Fig. 10.17 Structure of bisphenol A–epichlorohydrin polymer

Epoxy resins have been found to be of particular value in glueing metals together. For example, the shear strength of an epoxy resin (Araldite) between two mild steel sheets with an overlap ratio of 5 has been found to be as high as 48 MN/m^2 at atmospheric temperature, which is reduced to around 28 MN/m^2 at 150°C. It has been found that the use of epoxy glues is often to be preferred to other methods of metal jointing, as it does not weaken the metal structure.

Epoxy resins adhere tenaciously to many substances but cannot readily attach themselves to such plastics as polyethylene or vinyl resins.

Polysulphide and polyamide epoxy resins are used for flexible joints.

Several different epoxy resins are available and are processed by compression moulding, casting or laminating. The UTS of epoxy resins varies between 40 and 60 MN/m² and the compressive strength ranges from 70 to 110 MN/m². The dielectric strength is 18 kV/mm.

The heat deflection temperature depends markedly upon the type used. Flexible epoxy resins cannot be used above 80°C, while some of the heat-resistant types can be employed at temperatures up to 300°C without deflection taking place.

Epoxy resins are particularly suitable for glass-fibre reinforcement due to the fact that the material wets the glass well and that they can be cured with a very low degree of mould shrinkage. In this, epoxy resins contrast well with polyesters, which shrink badly during the curing process. Glass-fibre reinforcement can be carried out by filament winding, spraying upon a layer of chopped fibres, and laying by hand upon mats, cloth and woven materials. Pressure lamination can also be applied. Glass-fibre-reinforced epoxy resins have both compressive and tensile strengths of the order of 300 MN/m², even after many days of water immersion. The equilibrium water absorption of epoxy resins is low, varying normally from 0·8 to 2·0%.

Their resistance to dilute mineral acids is good, but to concentrated mineral acids and organic acids only fair. They have good resistances to alkalis, petrol and alcohol, while the resistance to ketones and chlorinated hydrocarbons is generally poor. Epoxy resins, however, withstand oils, greases, soaps and detergents well.

The main uses of epoxy resins in the building industry are in casings for electrical insulation, decorative panels, structural laminates, adhesives for bonding all types of materials and flooring finishes, either on their own or mixed with a filler.

POLYPROPYLENE

Polypropylene (see Fig. 10.18) is related to polyethylene, which will be dealt with in Chapter 11.

The over-all polymerization reaction is as follows:

$$n\text{CH}_2:\text{CH}.\text{CH}_3 \rightarrow (-\text{CH}.\text{CH}_3.\text{CH}_2-)_n$$

Polypropylene is made from propylene by using phosphoric acid and copper pyrophosphate as initiators.

Polypropylene has a softening-temperature of 160°C and is generally a good deal harder and tougher than polyethylene. It is cured mainly by casting, extrusion moulding and by blowing and has the remarkably low density of 0·9 kg/dm³. For moulding purposes it must be heated to between 230 and 280°C. Mould shrinkage is around 2%.

The material has excellent fatigue resistance and can be used at temperatures up to 120°C. It has extremely good electrical properties due to the fact that it has negligible water absorption (less than 0·01%).

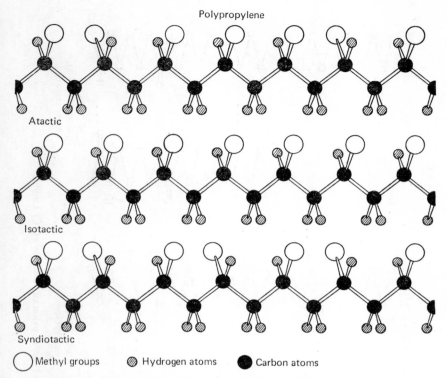

Fig. 10.18 Basic types of molecular arrangement of polypropylene: Atactic—random arrangement of methyl groups; Isotactic—all methyl groups on the same side; Syndiotactic —alternate arrangement of methyl groups
(By courtesy of ICI Limited)

Polypropylene is adversely affected by exposure to sunlight and should not be used on its own for external purposes, but only when mixed with carbon black or another kind of ultraviolet ray stabilizer. The chemical resistance of polypropylene against all common chemicals at room temperature is excellent, the material being hardly affected at all even by such drastic substances as ketones and chlorinated hydrocarbons. The chemical resistance, however, falls off drastically at higher temperatures and at 60°C polypropylene is badly affected by greases, oils, aromatic hydrocarbons, chlorinated hydro-carbons and ketones.

Uses

Polypropylene is mainly used for such purposes as hot water piping, for equipment which has to withstand steam temperatures, such as mould inserts used for casting concrete components, which are subsequently steam cured, and other similar purposes. Polypropylene is also employed in hospitals and laboratories as a steam, hot water and chemical resistant material.

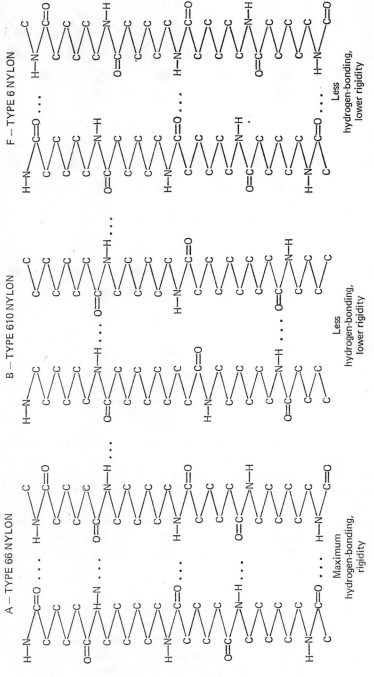

Fig. 10.19 Structures of various types of nylon (By courtesy of ICI Limited)

NYLON

Nylon is the name given to synthetic polyamide materials, where the polymerization process takes place by the reaction of a free carboxylic acid group with an amino group. In actual fact nylons closely resemble such natural products as horn, silk, hair, etc., which are based on polymerized amino-acids to form proteins (see Fig. 10.19).

Nylon 6.6

This type is made from hexamethylene diamine, $H_2N(CH_2)_6NH_2$, and adipic acid, $HOOC(CH_2)_4HOOC$, and is so-called because both the reactants contain six chain carbon atoms. The polymer is a simple chain one, produced by the reaction of the amine group at each end of the hexamethylene diamine with one of the acid groups of the adipic acid.

There are a number of other types of nylon in use which use different dibasic acids and diamines. Nylon 6.10 uses adipic acid and decamethylene-diamine, i.e. $NH_2(CH_2)_{10}NH_2$. It has a lower melting-point than nylon 6.6 and also has a lower moisture absorption.

The nylon grades with single figures are generally made from amino-acids, i.e. materials which have an amine group at one end and an acid group at the other. For example, nylon 11 is made from 1-amino-n-decanoic acid, $H_2N(CH_2)_{10}COOH$, which is polymerized to form continuous chains.

Nylon, when used as a casting material, is generally processed by extrusion, blowing and moulding and has excellent properties of durability and toughness.

An interesting property of nylon is its remarkably low coefficient of friction against many metals.

All grades of nylon have excellent long-term strength and very good impact strength. These, as well as many of the other physical and chemical functions of the material are, however, considerably modified when reinforcing agents, fillers and pigments are added. The material has excellent electrical properties due to its generally low water absorption figure, which is always well below 1%.

Nylon is classed from the fire-resistance point of view as self-extinguishing, although this property is modified considerably depending upon the additives employed.

Table 10.4 gives the main properties of the most common grades of nylon produced.

Uses

Nylon is used for a number of internal fittings, such as door and window furniture, and for purposes such as castors where use is made of the material's remarkably low coefficient of friction against most metals. Nylon is also employed for flexible tubing, electrical components and fittings.

o

TABLE 10.4

Properties	Nylon				
	6.6	6.10	6	11	12
Density, kg/dm³	1·14	1·08	1·14	1·04	1·01
Melting-point, °C	268	225	220	185	180
Maximum working temperature, °C	70—90	60—85	80—120 up to 160 if stabilised	60—100	80—90
UTS at 15°C, MN/m²	95	100	90	55	45
Mould shrinkage, %	1·5	1·5	1·0	1·2	0·8—2·0
Chemical resistance					
Water and steam	Fair	Fair	Fair	Fair	Good
	In all cases weatherability is improved by pigmentation				
Dilute mineral acids	Good	Good	Poor	Fair	Fair
Concentrated acids	Poor	Poor	Poor	Poor	Poor
Alkalies	Good	Good	Good	Very good	Fair
Alcohol	Fair	Fair	Good	Good	Good
Ketones, chlorinated hydrocarbons, esters	Fair	Fair	Good	Fair	Good
Petrol	Fair	Fair	Good	Good	Good
Oils, greases, detergents	Fairly good	Fairly good	Good	Very good	Very good

Literature Sources and Suggested Further Reading

1. BJORKSTEN, J., *Polyesters and Their Applications*, Reinhold, New York (1956)
2. BOENIG, H. V., *Unsaturated Polyesters*, Elsevier, Amsterdam (1964)
3. BOUNDY, R. H., and BOYER, R. F., *Styrene*, Reinhold, New York (1952)
4. BRENNER, W., LUM, D., and RILEY, M. W., *High Temperature Plastics*, Reinhold, New York (1962)
5. BRYDSON, J. A., *Plastics Materials*, Iliffe, London (1966)
6. DUFFIN, D. J., *Laminated Plastics*, Reinhold, New York (1966)
7. DU BOIS, H., and JOHN, F. W., *Plastics*, Reinhold, New York (1967)
8. FLOYD, D. E., *Polyamide Resins*, Reinhold, New York (1966)
9. GAYLAND, N. G. (editor), *High Polymers*, 15 volumes, Interscience, New York (1962–1968)
10. GOODMAN, I., and RHYS, J. A., *Polyesters*, 2 volumes, Iliffe, London (1965)
11. KORSHAK, V. V., and VINOGRADOVNA, S. V., *Polyesters*, Pergamon Press, Oxford (1965)
12. LEE, H., and NEVILLE, K., *Handbook of Epoxy Resins*, McGraw-Hill, New York (1967)
13. KINNEY, G. F., *Engineering Properties and Applications of Plastics*, Wiley, New York (1957)
14. MacTAGGART, E. F., and CHAMBERS, H. H., *Plastics and Building*, Pitman, London (1957)
15. McKELVEY, J. M., *Polymer Processing*, Wiley, New York (1962)
16. MEGSON, N. J. L., *Phenolic Resin Chemistry*, Butterworths, London (1958)
17. MORGAN, P. W., *Condensation Polymers*, Interscience, New York (1965)

18. OLEESKY, S., and MOHR, G., *Handbook of Reinforced Plastics*, Reinhold, New York (1964)

19. PATTON, T. C., *Alkyd Resin Technology*, Interscience, New York (1962)

20. PENN, W. S., *GRP Technology*, MacLaren, London (1966)

21. PENN, W. S., *PVC Technology*, MacLaren, London (1966)

22. 'Plastics, Material Guide', *Plastics Magazine*, November 1967

23. RITCHIE, P. D., *Physics of Plastics*, Iliffe, London (1965)

24. SCHENKEL, G., *Plastics Extrusion Technology*, Iliffe, London (1966)

25. SCHMITZ, J. V., *Testing of Polymers*, 3 volumes, Interscience, New York (1967)

26. SIMONDS, H. R., and CHURCH, J. M., *The Encyclopaedia of Basic Materials for Plastics*, Reinhold, New York (1967)

27. SKEIST, I., *Plastics in Building*, Reinhold, New York (1966)

28. SKEIST, I., *Epoxy Resins*, Reinhold, New York (1958)

29. VALE, C. P., and TAYLOR, W. G. K., *Amino Plastics*, Iliffe, London (1964)

30. YARSLEY, V. E., et al., *Cellulosic Plastics*, Plastics Institute, London (1964)

Chapter Eleven   Flexible Plastics Materials and Elastomers

11.1 Flexible plastics

POLYETHYLENE

Because of the very considerable stability of polyethylene and its cheapness, this material has been found to be one of the most useful stand-by substances in modern building technology. Not only is polyethylene widely used as a damp-proof membrane, when it is incorporated into foundation structures, but polyethylene foil is also commonly used as a vapour barrier for the protection of large structures during periods of inclement weather, etc. In the form of castings, blown and vacuum formed units, polyethylene is employed for such purposes as water tanks, housings, etc.

Polyethylene dates back to 1933 and was first produced on a commercial scale in the United Kingdom in 1939 and in the USA in 1943.

Ethylene gas, with a purity of 99·8%, is compressed to 140 MN/m² and is polymerized in the presence of certain initiators at this pressure and 200°C. It can also be made in the presence of benzene as a solvent using a temperature of 200°C and a pressure of 100 MN/m². The reaction is initiated by about 0·06—0·08% oxygen gas.

It can readily be expressed by the following equation:

$$n\mathrm{CH_2 : CH_2} \rightarrow (-\mathrm{CH_2 - CH_2 -})_n$$

The polymerization of ethylene is a very strongly exothermic reaction, the heat given off amounting to 92 kJ per monomer mole. After polymerization, the product obtained has a molecular weight of between 15,000 and 90,000. Nearly all the material is in the form of a sequence of methylene groups, i.e. polyethylene is simply a linear hydrocarbon with a very large number of $\mathrm{CH_2}$ groups. There are some branched units and there are also occasional ether (oxygen) bridges forming cross-linking. Polyethylene is made in two distinct grades, high density and low density, the high-density grade having a larger number of monomers in its polymer molecules.

Low-density polyethylene has a density of 0·92 kg/dm³ and high-density polyethylene a density of 0·965 kg/dm³. The melting-point of both grades is the same, namely 135°C, and both grades soften appreciably at temperatures above 100°C. The maximum service temperature of polyethylene should not generally exceed 75°C.

Polyethylene is formed by extrusion, casting, blowing, calendering and vacuum forming. The most usual and most useful form of the material in the building industry is sheeting, which accounts for over 60% of the total production of polyethylene. This is commonly made by extruding softened polyethylene through a circular slit and blowing continuously to form a seamless tube of thin-walled material. The processing temperature for extrusion is 180°C whereas for injection moulding, temperatures between 220 and 260°C are commonly employed. Mould shrinkage for low-density polyethylene is around 1·5% while with high-density polyethylene shrinkages may be as high as 3%. The coefficient of friction of polyethylene against many metals is low. For low-density polyethylene the coefficient of friction against aluminium or steel is 0·30, while for high-density polyethylene it may be as low as 0·20. Water has little effect, as polythene has a very low water absorption figure, which is of the order of 0·15% for the low-density variety and below 0·01 for high-density polyethylene.

The dielectric constant of both grades of polyethlyene is the same, namely 2·35 and the apparent volume resistivity is 10^{17} Ω.cm at 23°C, a value which is virtually unaltered at different relative humidities. Linear expansion is 12×10^{-5} degC^{-1} and refractive index is 1·54.

The gas permeability for the two grades of polyethylene is the following, given in m⁴/kN.sec:

	Low density	High density
Nitrogen	17×10^{-16}	5×10^{-16}
Oxygen	45×10^{-16}	15×10^{-16}
Carbon dioxide	240×10^{-16}	65×10^{-16}
Water vapour	1020×10^{-16}	300×10^{-16}

Chemical resistance is as follows:

	Low density	High density
Dilute mineral acids	Very good	Very good
Concentrated mineral acids	Good	Fair
Oxidizing acids	Poor	Poor
Alkalis	Good	Good
Alcohols	Good	Very good
Ketones	Fair	Very good
Aromatic hydrocarbons	Poor	Fair, but only for brief periods
Chlorinated hydrocarbons	Poor	Poor
Detergents	Fair	Good
Greases and oils	Good	Very good

When polyethylene is exposed to sunlight for long periods of time it is readily degraded; this is manifested by bad discoloration, crazing and a loss in mechanical strength. For this reason it is not advisable to use polyethylene for external purposes unless it has been stabilized by the addition of carbon black or a similar material.

The most usual form of polyethylene in the building industry is as sheeting, varying in thickness between 0·05 and 0·5 mm. This is coloured black whenever it is exposed to sunlight to prevent degradation. The main uses for such films are the following:

1 Wall and ceiling vapour barriers with thicknesses between 0·05 and 0·1 mm.
2 Damp-proof course membranes are placed between concrete slab and floor screed, and are up to 0·5 mm thick.
3 Ground cover in the case of sub-floor spaces to prevent rise of water vapour from damp ground.
4 Vapour barriers for cold storage rooms, for industrial built-up roofs, etc.
5 Steam tents are used for curing concrete sections.
6 Scaffold shelters enable work to proceed during periods of inclement weather.
7 Temporary glazing, covers for materials and equipment, dust and dirt screens.
8 As covers for soil slopes to prevent the earth from sliding down during periods of heavy rainfall.
9 As a bond breaker when using industrialized building systems such as the liftslab method or similar, where concrete is cast on top of concrete.

The physical properties of such polyethylene films are as follows:

Tensile strength	17 MN/m^2
Modulus of elasticity	158 MN/m^2
Elongation at breaking point	450%

Polyethylene becomes brittle at temperatures below $-100°C$. The below-ground longevity of polyethylene foil is greater than that of any competitive material, and the material is completely unaffected by water, salts or bacteria.

IRRADIATED POLYETHYLENE

When polyethylene is irradiated, cross-linking takes place between adjoining chains. This results in a material which keeps all the desired physical qualities of polyethylene such as flexibility, freedom from ageing, etc., but it increases the softening-temperature of the material by around 20 degC to about 80°C. The cross-linking takes place via double bonds in the polymers.

Polyethylene chains normally contain methylene groups as shown in Fig. 11.1.

Irradiation permits the polythene to be interlinked via such groups and produces materials stable up to 150°C and unaffected by solvents. The tensile strength of irradiated polyethylene is 13—14 MN/m² with a density of 0·92 kg/dm³.

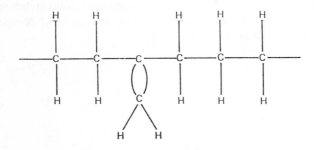

Fig. 11.1 Structure of irradiated polyethylene

CHLOROSULPHONATED POLYETHYLENE

This material is known under the trade name of 'Hypalon' (see Fig. 11.2) and is an elastomer, but with a chemical composition which differs completely from that of natural rubber.

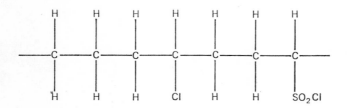

Fig. 11.2 Structure of hypalon (with random substitution)

The material is made by reacting chlorine and sulphur dioxide with polyethylene and producing a material containing about 1·5% sulphur and 27·5% chlorine in a polymer with approximate molecular weight of 20,000. Hypalon differs from other elastomers in that it is chemically saturated, i.e. it is completely unaffected by ozone, has excellent abrasion and weathering properties, as well as good oil and chemical resistance. The material has also good electrical properties and low moisture absorption. Hypalon is widely used in the building industry as a sealer for external uses, and as a cement. It is also employed as a roof covering, due to its ability to be calendered on top of plywood and other sheeting, and to form a good sealer at joints.

Hypalon needs no carbon black filler and because of this can be produced in a variety of colours.

PLASTICIZED VINYL COMPOUNDS

In contrast to unplasticized polyvinyl chloride and polyvinylidine chloride, which are rigid materials, there are several plasticized vinyl compounds that are very flexible (see Fig. 11.3).

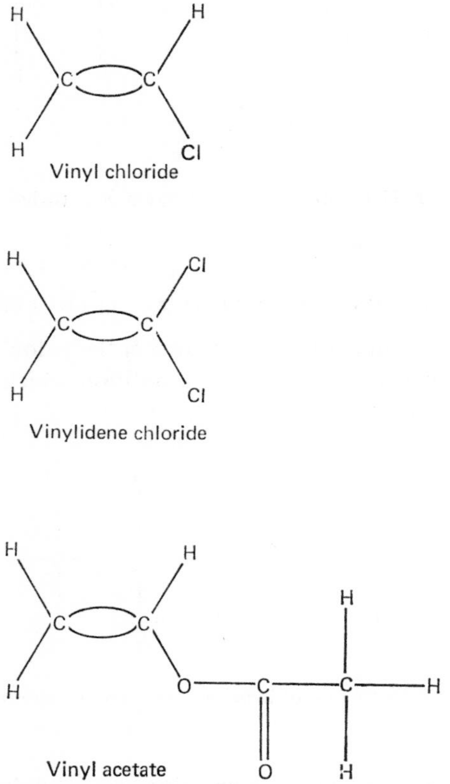

Fig. 11.3 Structures of vinyl chloride, vinylidene chloride and vinyl acetate

'Saran' is the trade name given to copolymers of these two materials containing 85% vinylidene chloride and 15% vinyl chloride. The plasticized form of Saran is a soft and flexible product which is widely used for roof flashing purposes. Saran has excellent chemical resistance to all common acids and alkalis with the exception of concentrated ammonia, is unaffected by water and also has negligible water vapour transmission properties. It behaves well against most organic solvents, but is attacked by halogens and amines.

Depending on the type of copolymer produced, it is possible to obtain a softening-temperature anywhere between 70 and 180°C.

Plasticized PVC has a density between 1·1 and 1·7 kg/dm³ depending upon type, and can be injection moulded, extruded or processed by vacuum moulding. The processing temperature is between 150 and 200°C. Mould shrinkage is, in general, between 1 and 5% depending on type. Plasticized PVC is extremely flexible at normal temperatures but at lower temperatures this falls off rapidly, requiring the additions of larger quantities of plasticizer to maintain the required flexibility. For example, the percentage plasticizer required to maintain the same flexibility at −20°C as existed at +10°C has to go up from 27 to 54%. Unlike polythene, PVC has a high coefficient of friction against metals and other materials as it possesses a tacky surface. The material has excellent stress cracking resistance and when used in the form of sheets has the following physical properties:

Tensile strength	13·7 MN/m²
Modulus of elasticity	11 MN/m²
Elongation at break	350%

The water vapour transmission figure is six times as high as that across an equivalent polyethylene film, but PVC sheeting has twice the resistance of the former to puncturing by nails and other sharp objects. The tear resistance of PVC is also better than that of polyethylene foil. Whereas polyethylene foil is stable down to a temperature of −100°C, PVC becomes brittle at temperatures below −24°C. PVC can be heated almost indefinitely at temperatures of around 100°C without suffering any heat ageing. Thermal expansion of PVC is 10^{-5} degC^{-1}. The material weathers reasonably well, but like polyethylene it should include stabilizers when exposed to ultraviolet light for long periods of time. Its resistance to ultraviolet light is, however, a good deal better than that of polyethylene. Under very wet conditions there is sometimes a danger that the plasticizer may be extracted, unless water insoluble types of plasticizers are used. Some plasticizers are also rather readily attacked by bacteria; this can be countered by the use of bactericides in the formulation or by the use of straight chain esters which are not subject to such an attack.

Plasticized PVC has an equilibrium water absorption between 0·5 and 1%. Chemical resistance is as follows:

Dilute mineral acids	Very good
Concentrated mineral acids and alkalis	Good
Benzene	Fair
Alcohols, ketones, chlorinated hydrocarbons	Poor
Soaps and detergents	Fair
Greases and oils	Poor

POLYVINYL ACETATE

The polymer is produced from vinyl acetate, a liquid which boils at 72°C, and polymerization is usually carried out in solution or as an emulsion. Redox types of activators are used to produce polymers with molecular weights of about 5,000 to 20,000. Vinyl acetates are mainly used as adhesives and coatings, but are also used as copolymers together with polyethylene for the production of plastic films.

The tensile strength of a 1:1 polyvinyl acetate–polyethylene copolymer is 21 MN/m², which is roughly 40% higher than that of PVC film and 25% higher than that of polyethylene. The material has a good modulus of elasticity, namely 75 MN/m², with an elongation of 600% at breaking point. Its resistance to tensile impact, puncture impact and tear resistance are all very much higher than those of either polyethylene film or PVC film. Puncture impact resistance is nearly three times that of equivalent polyethylene sheeting, while tear resistance is double.

Polyvinyl acetate–polyethylene has a softening-point of 68°C which is roughly the same as that of PVC.

11.2 Rubber

Rubber is used increasingly in the building industry for such purposes as flooring, anti-acoustic bearings, flexible joints, tubing, weathering, seals and flashing and many other purposes.

The term *latex* is used in the rubber industry to describe naturally occurring rubber dispersions which are obtained as secretions from a number of different plants. The most important of these bears the biological name: *Hevea Braziliensis*. However, during the last 20 years or so the importance of natural rubber has been overshadowed by the development of a large number of synthetic rubbers. Natural rubber, however, still has an important part to play in the rubber industry.

NATURAL RUBBER

Rubber latex, which is obtained directly from the plant by making insertions in the bark, is coagulated with acetic or formic acid to produce the crepe rubber of commerce. Natural rubber latex has the following composition:

Water	55%
Rubber hydrocarbons	35%
Proteins	4·5%
Acetone extracts	4%
Impurities such as inorganic salts, amino acids, etc.	1·5%

The rubber hydrocarbons consists almost exclusively of polymers of the material isoprene (see Fig. 11.4), which forms both *cis*- and *trans*-polymeric molecules (see Fig. 11.5). The average molecular weight of natural rubber is between 750,000 and 2,500,000, and the material is in the form of long chains, which are intertwined with each other in the form of coils. When rubber is stretched, the molecules uncoil with the single bonds in the structure rotating

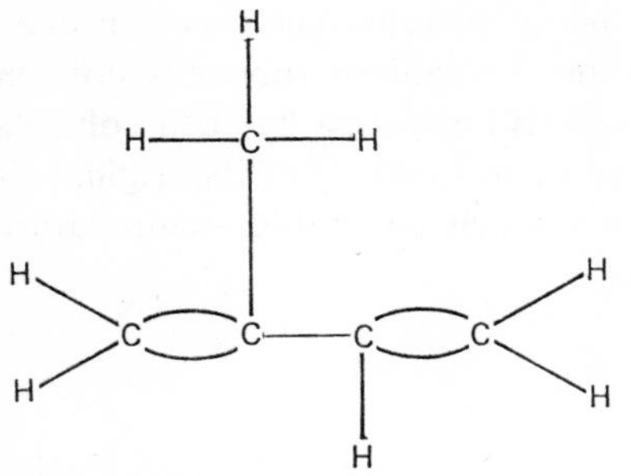

Fig. 11.4 Structure of isoprene

as the chains straighten out. But this position is unstable thermodynamically, and when the restraint is ended, the rubber polymers return to their original coiled shape.

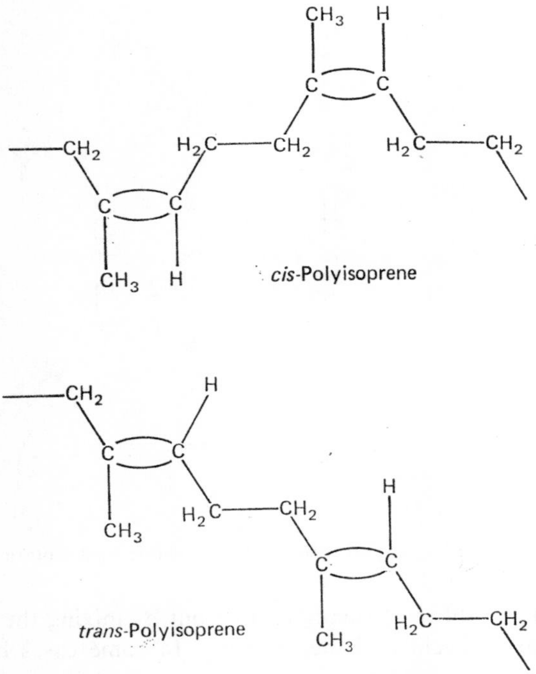

cis-Polyisoprene

trans-Polyisoprene

Fig. 11.5 Structures of *cis*- and *trans*-polyisoprene (the former being the arrangement in natural rubber)

VULCANIZATION OF RUBBER

Vulcanization of rubber radically changes the nature of the material. It very much increases the tensile strength and resistance to abrasion, and also reduces the permanent set and compression set of the material. Unvulcanized rubber is soluble in many solvents such as benzene, chloroform, carbon tetrachloride and hexane. Vulcanized rubber is not readily soluble in these materials, although it is still attacked by such solvents and swells when in permanent contact with them to 300% of its original size and more. Finally, vulcanized rubber has a much better high-temperature resistance than un-vulcanized rubber.

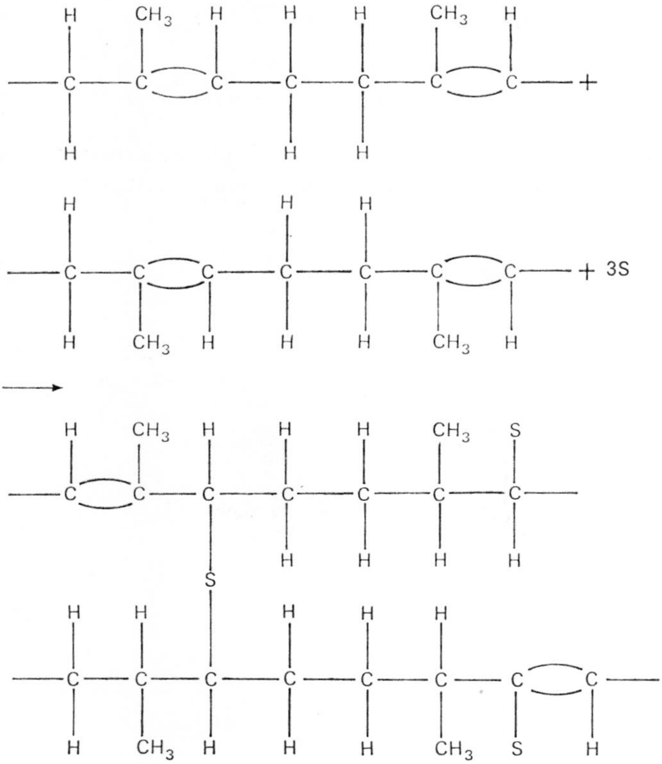

Fig. 11.6   Vulcanization with sulphur (possible representation)

Vulcanization of rubber is usually carried out by mixing the crude rubber with a vulcanization agent and then heating. In some cases heating is not necessary because the vulcanization agent causes initiation at ordinary temperatures. There are even methods of carrying out the vulcanization of rubber under the action of gamma rays only.

The main vulcanization agents commonly used are the following: sulphur (see Fig. 11.6), hydrogen sulphide, sulphur dioxide, sulphur chlorides, benzoyl-peroxide, diazoaminobenzene $C_6H_5.N:N.NH.C_6H_5$, tetra-alkylthiuram di-sulphides (see Fig. 11.7), quinones, aniline phenols, as well as some inorganic materials such as mercuric oxide, selenium, tellurium and sulphur thiocyanate.

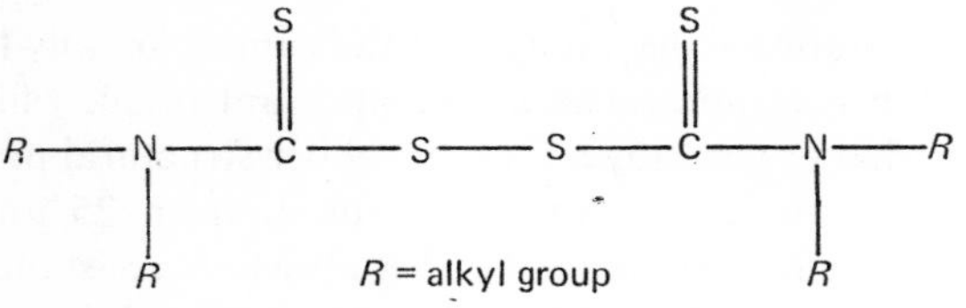

Fig. 11.7 Structure of tetra-alkylthiuram disulphide

When normal vulcanization is practised, the time taken may be very long, especially when large rubber objects are concerned. This is very much reduced when so-called *accelerators* are employed. Up to 50 different organic compounds are widely used to cut down vulcanization times. The most useful is mercaptobenzothiazole (see Fig. 11.8) and its derivatives. This material is commonly known as MBT and works best at a temperature of 110°C and above.

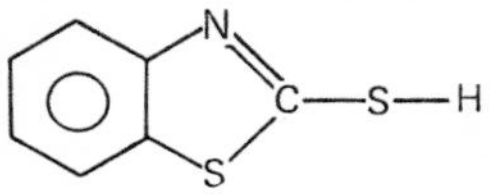

Fig. 11.8 Structure of 2-mercaptobenzothiazole

Accelerators reduce vulcanization times from a matter of hours to a matter of minutes, and often also enable temperatures to be reduced drastically. Without an accelerator between 8 and 10% sulphur addition to natural rubber is required; this can be reduced to 1 to 3% if MBT or a similar material is added. Accelerators usually also need the addition of zinc oxide to the rubber, to enable the formation of a zinc salt, an essential part of the chemical reaction. Sometimes, on the other hand, it is necessary to slow down the vulcanization process. In such circumstances retarders such as acetylsalicylic acid (see Fig. 11.9) are added.

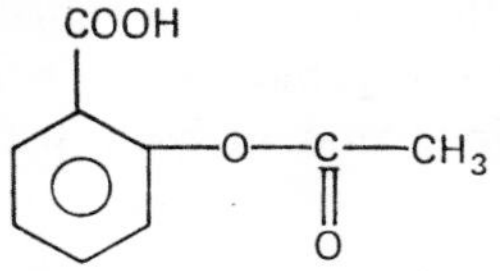

Fig. 11.9 Structure of acetylsalicylic acid (aspirin)

OTHER COMPOUNDING INGREDIENTS

In addition to vulcanization agents and accelerators, the following materials are usually added to the rubber mix prior to vulcanization.

Fillers

These are added to improve the strength of the rubber, mainly from the point of view of abrasion resistance. The most important of such fillers is carbon black, which is added to virtually all heavy rubber structural parts used in the building industry. Various size ranges are used, from 25 μm diameter to 400 μm. Small particle sizes improve the abrasion resistance and tensile strength, but lower the resilience of rubber. The main white reinforcing filler used is zinc oxide, while the main extending fillers which do not affect the colour of the rubber are calcium carbonate, barium sulphate and china clay.

Softening agents

Stearic acid $C_{17}H_{35}COOH$ and oleic acid $C_{17}H_{33}COOH$ are the most usual softening agents and help in the dispersion of fillers. Other materials commonly employed are petroleum jelly, paraffin wax and various types of resins. The main function of the softeners is to increase the degree of tack of the un-vulcanized product and to help the vulcanization process by preventing the sulphur from coming out of solution.

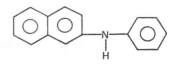

Fig. 11.10   Structure of phenyl-β-naphthylamine

Protective agents

Rubber is very vulnerable to oxidation, particularly if free ozone is present in the atmosphere. Attack by oxygen means that rubber begins to crack and also often starts to become tacky. This is commonly known as *perishing*. This attack by oxygen is accelerated by sunlight, heat and the presence of oils and solvents; certain materials called antioxidants, such as phenyl-β-naphthyla-mine (PBN) (see Fig. 11.10) reduce the tendency of rubber to perish due to oxidation. Only very small quantities of such antioxidants are usually necessary. Colouring matter is often mixed with rubber. Apart from red oxide of iron, most colouring matter is organic. Strong white colours are obtained by using titanium dioxide or zinc oxide, or both, with the rubber.

SYNTHETIC RUBBERS

A large number of synthetic rubbers are on the market; these differ chemically from natural rubber, but can in most cases be mixed with natural rubber and with each other. In fact, copolymerization of various types of rubber ensures that the properties of the final product are as close as possible to the ones wanted.

The main types of synthetic rubber and their ASTM (American Society for Testing and Materials) abbreviations are the following:

1 Butadiene rubbers (BR).
2 Synthetic isoprene rubbers (IR).
3 Chloroprene rubbers (CR).
4 Acrylate–butadiene rubbers (ABR).
5 Isobutylene–isoprene rubbers (IIR).
6 Nitrile–butadiene rubbers (NBR).
7 Nitrile–chloroprene rubbers (NCR).
8 Pyridine–butadiene rubbers (PBR).
9 Styrene–butadiene rubbers (SBR).
10 Styrene–chloroprene rubbers (SCR).
11 Styrene–isoprene rubbers (SIR).
12 Silicone rubbers.

Only a few of these rubbers are commonly used in the building industry and for this reason descriptions are restricted to these.

The most common of all synthetic rubbers is the *butadiene–styrene* (see Figs. 11.11 and 11.12) copolymer, which is used for many general purposes.

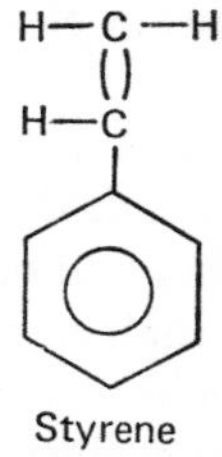

Fig. 11.11   Structure of styrene

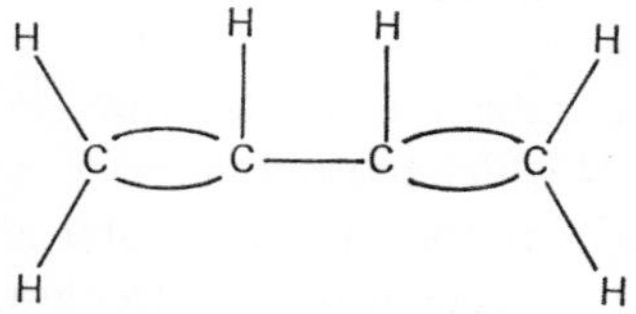

Fig. 11.12   Structure of butadiene

The two liquid monomers are mixed in the proportion of 25% styrene to 75% butadiene and are emulsified, prior to primary curing. After unreacted styrene and butadiene have been removed, the SBR latex is coagulated with dilute sulphuric acid and salt solution. SBR has a poorer tack than natural rubber and also an inferior tensile strength and poorer resilience. Its resistance to abrasion is greater than that of natural rubber, and it has been found better than natural rubber with respect to skid resistance. For this reason SBR is widely used in the building industry for floor tiles and similar purposes. It has considerable advantages over natural rubber when used for this purpose because it is less subject to ageing, discoloration and has greater resistance to wax polishes. SBR is more difficult to cure than natural rubber and therefore needs more accelerator.

Neoprene is the name given to the various elastomers made from chlorine substituted isoprenes such as 2-chloro-1,3-butadiene (see Fig. 11.13) which is

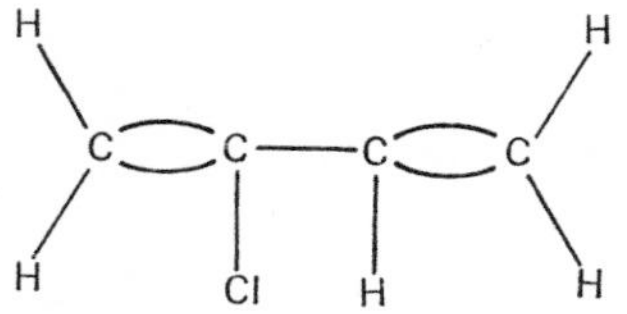

Fig. 11.13   Structure of 2-chloro-1,3-butadiene

called chloroprene. Chloroprene is suspended in water to form an emulsion and, after partial polymerization, is coagulated with acetic acid. Vulcanization takes place in the normal way. Pure neoprene has a tensile strength of 28 MN/m² which compares well with that of natural rubber, with an elongation at break of 900%. The modulus of elasticity at 600% elongation is equal to 7 MN/m².

Neoprenes are much more stable against oxidation than natural rubber, and only swell moderately in oils and chemicals. The gas permeability of neoprene products is lower than that of natural rubber. Another feature of neoprene is its resistance to sunlight, which enables neoprene to be used for external purposes in construction. Neoprene is widely used in the building industry for such purposes as flashing, rainwater exclusion, etc., purposes for which neither natural rubber nor SBR would be too satisfactory, because both these rubbers are liable to perishing. Neoprene, unlike most rubbers, is highly flame-resistant and self-extinguishing.

Acrylonitrile–butadiene rubber is made from 75% butadiene and 25% acrylonitrile $CH_2:CH-CN$ which are polymerized partially in the emulsified state and coagulated. Nitrile rubber has low water absorption, a high resistance to abrasion, as well as excellent tensile strength, elongation, tear resistance and chemical resistance.

It is totally unaffected by solvents such as hexane, paraffin, or by lubricating oil and water. It suffers slight swelling when immersed in ethanol and petrol and is very badly affected by such solvents as acetone, benzene and carbon tetrachloride. Nitrile rubbers can be bonded easily to steel with phenolic resins.

In the building and construction industries considerable use is made of these bonded products for such purposes as prefabricated roofing sections and sidings. The material is also used for purposes where rubber has to be exposed continuously to oils and solvents.

11.3 Manufacturing processes used in the rubber industry

THE USE OF RUBBER LATEX

Natural rubber latex is stabilized by the addition of certain preservatives and is used as the raw material for the manufacture of many materials. Latex foam sponge, which is widely used for the manufacture of upholstery, is made directly from rubber latex. The material is concentrated so that its rubber content is 60%, and the necessary compounding ingredients are all added in the form of water dispersions. The mix is then converted into a foam of the consistency of whipped cream. A gelling agent is added, followed by pouring into a mould and curing at 100°C. The article is then washed in water and dried.

THE USE OF RUBBER FOR FLOORS, TILING AND ROAD SURFACES

For flooring, mixes are made up in which the percentage rubber used seldom exceeds 25%, the rest being made up by various fillers to limit the elasticity of the rubber and to reduce creep. Marbled and mottled effects are produced by using runners of different coloured rubbers, which are calendered together. Vulcanization is carried out in large presses or continuously by means of a Rotocure machine. Tiles are made in the same way as sheet flooring, except that they are cut afterwards from vulcanized sheet.

Rubber latex is often combined with Portland cement, cork, sawdust, sand and other materials to produce flooring, which adheres well to other surfaces and is very abrasion-resistant. For road construction rubber is mixed with tar and bitumen in powder form, together with pieces of ready-vulcanized chips of rubber. Such surfaces are very durable and skid-resistant.

RUBBER SOLUTIONS

Rubber solutions are solutions obtained from uncured rubber in various solvents. The most common are petroleum fractions, benzene, carbon tetrachloride and chloroform. Carbon disulphide, which used to be employed for this purpose, is now illegal due to its toxicity. The main uses for rubber solutions are in floor laying, for other operations involving the sticking of rubber and plastics, and for the proofing of fabrics. The solvents used govern the speed of initial drying of the rubber cement. The lower the boiling-point the faster is the rate of evaporation. The chlorinated hydrocarbon solvents are non-flammable, yet highly toxic, which means that when they are used it is essential to practise good ventilation. The petroleum fractions and benzene, toluene and xylene solvents are less toxic but rather inflammable. When liquids with a flash-point below 25°C are used, the material is termed

P

'highly inflammable' and special laws concerning the handling of such materials apply.

Most forms of rubber bond readily to metals with considerable tenacity. Such bonds are due to direct chemical forces. When vulcanized rubber is bonded to copper, for example, a reaction takes place between the sulphur of the rubber and the copper, producing co-ordinated S/Cu bonding (see Fig. 11.14). Neoprene will also bond to copper or brass, even in the absence of

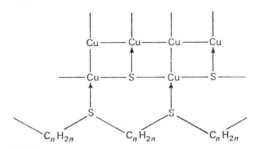

Fig. 11.14   Bonding of vulcanized rubber to copper

sulphur, but in such a case the bond is due to polar forces only. When rubber is to be bonded to metals other than copper it is usually necessary to apply a layer of ebonite to the surface of the metal by means of a roller, or in the form of a solution. Rubber can then be bonded to the ebonite by cross-linking. The strength of such metal/rubber bonds is unaffected by low temperatures but is weakened when temperatures are increased. Bonds between non-copper metals and ebonites made from SBR or ABR are much stronger than when natural rubber is used in the ebonite. Ebonite, which is simply rubber with a very high percentage of sulphur (cross-linkages up to 30%), cannot be employed to make a bond with copper or brass, as it forms a powdery interlayer of copper sulphide, which does not bond readily to the rubber.

Another form of rubber which bonds readily to metals is chlorinated rubber (see Fig. 11.15), produced by the reaction of hydrogen chloride with rubber dissolved in benzene. The chlorine content of this form of rubber can

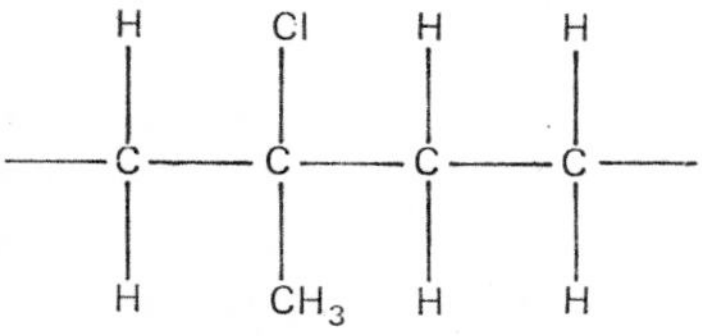

Fig. 11.15   Partially chlorinated rubber

go as high as 68% (see Fig. 11.16). Such chlorinated rubber is used as an intermediate layer to permit neoprene and nitrile rubbers to be bonded to metals. Chlorinated rubber cannot be used directly with natural rubber or SBR, because of incompatibility. Interlayers of neoprene are required in such cases. Rubbers can also be bonded to metals using various cements. One of

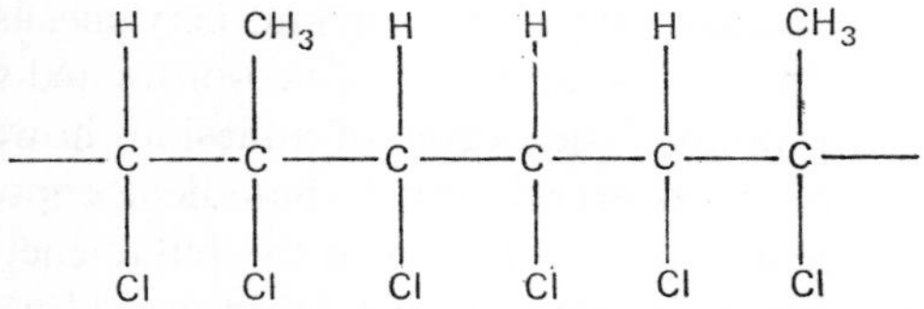

Fig. 11.16   Fully chlorinated rubber (68% by weight)

these is a solution of a complex isocyanate in methylene chloride (see Fig. 11.17), which forms a bond which is highly resistant to oils. Phenol formaldehyde resins may also be used to bond vulcanized neoprene and nitrile rubbers to metals.

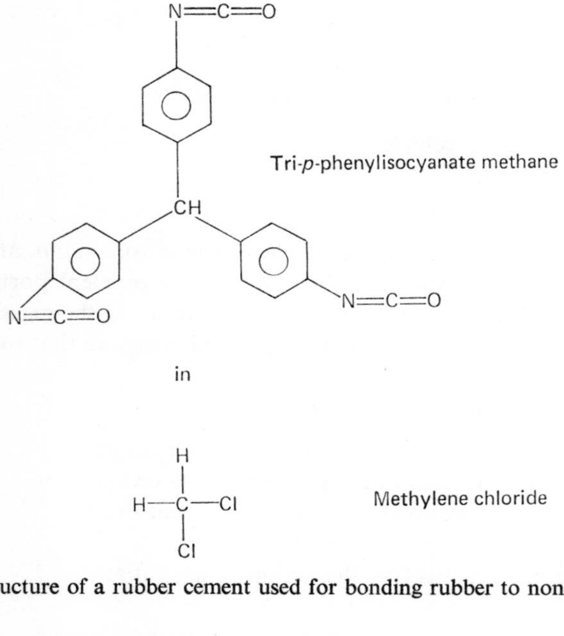

Tri-*p*-phenylisocyanate methane

in

Methylene chloride

Fig. 11.17   Structure of a rubber cement used for bonding rubber to non-cuprous metals

11.4 Textiles used in conjunction with rubber

Some 80% of all rubber goods produced use rubber in conjunction with textiles. In the building industry rubber–textile products are used for purposes connected with concrete casting, for driving belts, conveyor belts and for flexible joints.

P*

The main reason why rubber is often compounded with woven materials is that the tensile strength of even the toughest rubber is poor, a mere 27 MN/m² as against up to 700 MN/m² for some textiles. On the other hand, most textiles break at an elongation of 10% while rubbers can show elongations of up to 600% and more without fracture.

The following woven fabrics are commonly used in conjunction with rubber processing: wool, silk, cotton, linen, rayon, nylon, Terylene, asbestos and glass fibre weaves. All these materials can be readily co-ordinated with all forms of rubber, producing firm bonds. The degree of adhesion, however, varies with the nature of the rubber used. Acrylonitrile–butadiene copolymers have the highest cohesion to most fibres, while that of the butadiene – styrene type is least. Adhesion to nylon is less than that of other fibres for all rubbers, due to the smooth surface of nylon. The degree of adhesion to fabric also depends largely upon the type of weave used, twill weaving being one of the best.

Rubber is applied to textiles either by impregnation with a mixture of rubber and resorcinol formaldehyde resin, or by impregnating the fibres first with the resorcinol formaldehyde, followed by the attachment of the rubber coating. Chlorinated rubber or polyisocyanate are also used as fixing agents.

11.5 Sealants

Sealants consist in general of three components:

1 Basic non-volatile vehicle.
2 Pigment portion.
3 Solvent.

The non-volatile vehicle is often an elastomer of some kind, although more primitive sealants usually employ linseed oil. The pigment portion serves to control the flow and also to impart opacity or colour to the mass. The solvent is added for the sole purpose of reducing the viscosity so that the sealant can be easily applied.

Sealants are basically of five types:

1 Where the vehicle oxidizes and thickens as a result.
2 Where the sealant does not dry and remains tacky.
3 Where setting is due to evaporation of solvent only.
4 Where the sealant is set by high temperature.
5 Where the sealant is set by the action of an initiator.

OXIDIZING SEALANTS

These are mostly made from vegetable oils and similar materials, and contain between 70 and 80% solid by volume. These materials are the cheapest of all, but have the disadvantage of poor elasticity. The percentage elongation at break may be as low as 25%. Linseed oil putty skins overnight and sets firm after 4—8 weeks. Its adhesion is only fair, and it can seal gaps up to 2·5 cm wide. Life expectancy is lower than for other kinds of sealants, being a maximum of around 8 years. The material cracks as it ages.

NON-DRYING SEALANTS

These are commonly based on polybutene or polyisobutylene, employing a stabilizer, pigments and fibrous materials. Solvents are not used. Elongation at break is roughly the same as that of the ordinary vegetable oil putty, but their life expectancy is up to 15 years. Unlike vegetable oil putties, there is little loss of adhesion or hardening with age. Cost is virtually the same as that of vegetable oil putty.

SOLVENT EVAPORATION TYPES OF SEALANTS

Two types of these are produced, one based upon butyl rubber and the other upon acrylic types of rubber. They are heavy gums with between 70 and 90% solid content and are available in all colours. They are usually applied by a gun. Elongation at break of both is between 200 and 600% and they are mainly used for dealing with small cracks below 1 cm in width. These sealants skin in about 2—4 hr and set firm in 4—6 weeks. Acrylic sealants last longer than butyl sealants—20 years against 10 years. The disadvantage of acrylic sealants is, however, that they must be heated for application. Butyl sealants remain somewhat tacky throughout their life. Cost is high, especially of acrylic sealants.

THERMOSETTING SEALANTS

These are usually based on various vinyl copolymers, which are mixed with plasticizers and fillers. Some solvent is often also present, but this seldom exceeds 20%. These sealants can be applied by means of a spatula or a gun, and the degree of adhesion varies considerably according to formulation. The material is cured by heating for 15—60 min at 120—180°C. As the sealant ages it hardens and loses adhesion. The life expectancy is no higher than that of ordinary linseed oil putty.

CHEMICAL CURING SEALANTS

These are based on polysulphide, silicone or polyurethane polymers, and contain fillers and resins as well. These sealants usually have no solvents and are commonly applied by gun. They have very good adhesion and can be obtained in most colours. Elongation at break is up to 600% and they set rapidly, provided the temperature is favourable. For example, at 5°C such sealants need 6 days to set, while at 20°C this period is reduced to 15 hr and at 40°C to between 1·5 and 4 hr. The material has a life of up to 20 years. Its main disadvantage is price. Some grades may cost more than ten times as much as vegetable oil putties.

11.6 Bitumen

Bitumen, or asphalt as it is called in the United States, is the residue obtained from the distillation of crude oil. In Great Britain the term asphalt is usually reserved for mixtures of bitumen with inert mineral matter.

Some bitumen is also found naturally either alone or mixed with mineral matter. The material has a variable hardness depending upon the quantity of light fractions removed during the distillation process. Its softening-point is between 25 and 170°C, depending on grade.

Bitumen can be oxidized by bubbling air through the residue during manufacture. This produces a product with some rubbery properties, which does not soften as readily on heating or become as brittle on cooling as ordinary grades of bitumen.

The main use of bitumen is in road construction, where it is mixed with mineral matter to form the well-known tar—macadam. Bitumen is often used for such purposes in the form of emulsions, carrying 50—65% bitumen suspended in water.

Bitumen is often also used as an anti-corrosive protective coating, and is usually applied by dissolving the molten bitumen in a solvent in a closed vessel. These coatings are then applied by brush and spray. Two coats normally last one year. Thicker coatings of bitumen (in excess of $\frac{1}{2}$ mm) are usually applied by dipping or trowelling of the molten bitumen itself. Bitumen has good chemical resistance to acids and bases and most inorganic salt solutions. Its resistance to organic solvents of any kind is, however, virtually nil.

Bitumen has good adhesive properties and it is used in the building industry for sticking flooring tiles and roofing tiles, for the protection of concrete and brickwork against water, and coordinated with other materials for such varied purposes as bituminized waterproof paper, bitumen-lined piping, bituminized felt, roof shingles, etc. Such treatment is usually carried out by impregnating the porous carrier material with molten bitumen by dipping, followed by passage through rollers to remove the excess material.

Coated felt is usually dusted afterwards with talcum powder to prevent it from sticking to adjacent surfaces during transportation.

Literature Sources and Suggested Further Reading

1. ABRAHAM, H., *Asphalts and Allied Substances*, 5 volumes, Van Nostrand, New York (1960–1963)

2. ALLIGER, G., and SJOTHUN, I. J., *Vulcanisation of Elastomers*, Reinhold, New York (1964)

3. BATEMAN, L., *The Chemistry and Physics of Rubberlike Substances*, MacLaren, London (1963)

4. BOENIG, H. V., *Polyolefines*, Elsevier, Amsterdam (1966)

5. DAVEY, A. B., and PAYNE, E. W., *Rubber in Engineering Practice*, MacLaren, London (1964)

6. HOFMAN, W., *Vulcanization and Vulcanizing Agents*, MacLaren, London (1967)

7. CHEVASSUS, F., and BROUTELLES, R. DE, *The Stabilization of Polyvinyl Chloride*, Arnold, London (1963)

8. KRAUS, G., *Reinforcement of Elastomers*, Interscience, New York (1965)

9. HOIBERG, A. E., *Bituminous Materials*, 3 volumes, Interscience, New York (1964–1966)

10. JAMES, D. I., *Abrasion of Rubber*, MacLaren, London (1967)

11. MCPHERSON, A. T., and KLEMIN, A., *Engineering Uses of Rubber*, Reinhold, New York (1956)

12. MORTON, M., *Introduction to Rubber Technology*, Reinhold, New York (1959)

13. NAUNTON, W. J. S., *The Applied Science of Rubber*, Arnold, London (1961)

14. RAFF, R. A. V., and ALLISON, J. B., *Polyethylene*, Interscience, New York (1956)

15. RENFREW, A., and MORGAN, P., *Polythene*, Iliffe, London (1957)

16. SAUNDERS, J. H., and FRISCH, K. C., *Polyurethanes*, Interscience, New York (1964)

17. SCOTT, J. R., *Physical Testing of Rubbers*, MacLaren, London (1965)

18. STERN, H. J., *Rubber – Natural and Synthetic*, MacLaren, London (1967)

19. WHITBY, G. S., *Synthetic Rubber*, Wiley, New York (1954)

20. TRAXLER, R. N., *Asphalt*, Reinhold, New York (1961)

21. WILSON, B. J., *British Compounding Ingredients for Rubber*, Heffer, Cambridge (1964)

22. ICI LIMITED, Technical Brochures

Chapter Twelve   Cellular Plastics

The development of cellular plastics has revolutionalized methods of insulating buildings. These materials are normally far superior to competing substances and also show a decided cost advantage. Cellular plastics fall into two categories: (*a*) those with closed pores, and (*b*) those with open pores. Closed-pore types can double as damp-proof-course materials, but their failure to permit water vapour to pass can sometimes cause trouble. The open-pore grades generally constitute a good barrier to liquid water, due to the negative capillary effect of most plastic materials, but allow air and water vapour to pass without hindrance. Vapour barriers are necessary in such a case to prevent condensation of water within the pore structure.

THERMAL CONDUCTIVITY OF CELLULAR PLASTICS

Since in all the cellular plastics the volume of solid material used is very small in comparison to the amount of air enclosed, the thermal conductivity is virtually that of completely still air. Exceptions only occur with very dense cellular plastics, where the volume of material can no longer be neglected, and very lightweight samples where there is some radiant heat transfer through the no longer opaque mass. With normal cellular plastics filled with air, the thermal conductivity is practically constant at 0·035 W/m.degC at 20°C. Thermal conductivities naturally vary with temperature. The thermal conductivities of closed-pore plastic foams which are filled with gases other than air differ, however. The CO_2-filled foams have thermal conductivities of 0·039 W/m.degC while the Freon-filled types have conductivities as low as 0·025 W/m.degC (see Fig. 12.1). Several other gases are used as blowing agents, each with their own thermal conductivity figures.

The main groups of cellular plastics used in the building industry are the following:

1 Vinyl foams.
2 Expanded polystyrene.

3 Expanded ebonite and rubber.
4 Phenolic foam.
5 Foamed polyethylene and related materials.
6 Polyurethane foams.
7 Urea-formaldehyde foam.

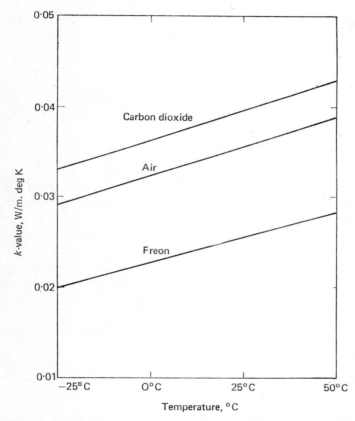

Fig. 12.1   Relationship between the thermal conductivity of foams and
the ambient temperature

12.1 Vinyl foams

These are made from mixtures of vinyl acetate, vinyl chloride and plasticizers.
A typical formulation would be:

Vinyl chloride	85 parts
Vinyl acetate	15 parts
Tricresyl phosphate	138 parts
Dioctyl phthalate	12 parts
Aluminium stearate	1·5 parts

The mix is placed in a closed container and subjected to a pressure of between 7 and 55 bar of either carbon dioxide or a chlorofluorohydrocarbon. The resin mix absorbs the gas, the absorption process being helped by a number of baffles within the vessel. The foam is produced when the pressure is released and the material is then cured to a final density as low as 32—48 kg/m³.

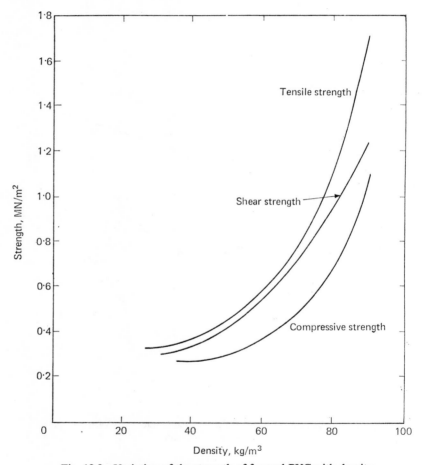

Fig. 12.2 Variation of the strength of foamed PVC with density

Vinyl sponge is made by dissolving some hydrazine compounds, which give off nitrogen on heating in the plasticizer – resin wax. This material has a somewhat different structure to vinyl foam, and possesses the following mechanical properties, depending upon density (see Fig. 12.2):

Tensile strength	0·5—2·4 MN/m²
Compressive strength	0·3—2·0 MN/m²
Maximum operating temperature	45—65°C

Expanded PVC is particularly suitable for use where its very considerable compressive strength and the non-friable nature of its surface can be made use of. Due to its rigidity, it is possible to produce large prefabricated panels from PVC foam, which can be sandwiched with other materials. A typical sandwich panel containing a core thickness of 7·5 cm and clad on both sides by 20 s.w.g. light alloy has been tested as follows.

The panel, measuring 185 cm × 30 cm, is supported at its ends and a load of 90 kg is applied at its centre. The central deflection measured amounted to only 1 cm. Similar deflections were found when the panels were covered on both sides by thin sheets of asbestos cement or plywood.

Vinyl foam does not support combustion and is extremely resistant to chemical attack. It has a 7 days' water absorption figure of between 3·0 and 3·8% and a water vapour diffusance rate of between 0·01 and 0·02 kg/m². hr. bar for board 1 in (25·4 mm) thick. This very low figure of water vapour transmission enables PVC foam to be used as its own water vapour barrier.

The material is chemically resistant to aliphatic hydrocarbons, dilute acids and alkalis, alcohol, carbon tetrachloride and sea water. It is attacked by acetone, ethyl acetate, ethylene dichloride and aromatic hydrocarbons.

The material is used as permanent shuttering to produce a concrete surface backed by a thermal insulation layer, for the construction of insulated ground floor slabs, for highly insulated cladding panels and for roof sections.

The vinyl foams are by far the strongest of the cellular plastics made. Their main properties are summarized in Table 12.1. The material can thus be used for purposes where competing substances such as expanded polystyrene would be too weak.

TABLE 12.1

Density, kg/m³	35	48	60	90
Cell structure	Coarse	Fine	Fine	Coarse
Compressive strength, kN/m²	275	345	690	1,240
Compressive modulus, MN/m²	12	12	24	28
Shear strength, kN/m²	310	520	690	1,030
Shear modulus, MN/m²	12	8·3	16	28
Flexural strength, kN/m²	345	550	970	1,380
Tensile strength, kN/m²	345	550	1,380	1,730
Coefficient of linear expansion, $10^{-5}degC^{-1}$	3·6	4·7	5·3	4·3

12.2 Expanded polystyrene

This material is probably the most popular high-void insulator used today, due to its very wide applicability and comparatively low price.

The raw material is supplied in beads which are between 0·4 and 0·8 mm in diameter, and which have a bulk density of roughly 1,040 kg/m³. The beads, which are either white or clear, are soaked in an expanding agent (usually butane) and are delivered in sealed drums. For low-density expanded polystyrene, steam heating is used, while hot water pre-expansion is employed for the production of denser substances. In all cases some pre-expansion is required. Moulding proceeds at about 100—110°C.

During both the pre-expansion and the final expansion, the globules are melted and at the same time the butane is evaporated, causing them to swell. The final state is a board or other shaped form in which the individual expanded globules have coalesced. The usual density of expanded polystyrene as used in the building industry varies between 16 and 36 kg/m³. An alternative method of producing expanded polystyrene is to melt polystyrene and to mix it with a blowing agent, followed by extrusion through an orifice. The expanded polystyrene made in this way has a density of 35 kg/m³ and does not possess the usual bead structure. It is manufactured under the trade name 'Roofmate'. The properties of the most usual forms of expanded polystyrene are given in Table 12.2.

TABLE 12.2

	Moulded polystyrene			Blown polystyrene
Density, kg/m³	16	20	24	35
Compressive strength, kN/m²	110	130	180	250
Tensile strength, kN/m²	200	230	300	410
Shear strength, kN/m²	140	180	220	330
Water vapour transmission, kg/m².hr.bar, for 1-in thick (25·4 mm) board	0·05	0·04	0·035	0·03
Coefficient of linear thermal expansion, all grades, degC⁻¹		$6·5 \times 10^{-5}$		

The relationship between the density and thermal conductivity of expanded polystyrene at 20°C is shown in Fig. 12.3.

All forms of expanded polystyrene have a very low degree of water absorption, which varies according to grade between 2 and 3% by volume after 7 days' water immersion for the moulded type, and about 1% for the denser grades. When ignited, all forms of expanded polystyrene soften and collapse. It is, however, possible to modify the combustion characteristics of expanded polystyrene by incorporating certain flame-retardent materials in the mix. The maximum service temperature is 75°C and softening takes place at 90°C. The compression and flexural modulus of the 16 kg/m³ grades are 4·8 MN/m² and 9·3 MN/m² respectively (see Figs. 12.4, 12.5 and 12.6).

Coloured expanded polystyrene is made by colouring the beads before use. When moulding thick blocks it is sometimes difficult to get completely uniform density and variations of up to 10% are common. The blown type, on the other hand, is usually extremely uniform in density. Expanded polystyrene is readily attacked by many solvents, and for this reason considerable care must be taken when painting it or glueing the material.

Expanded polystyrene is used for underfloor insulation, when it acts as a permanent shuttering for concrete which is cast on top. It is impracticable to

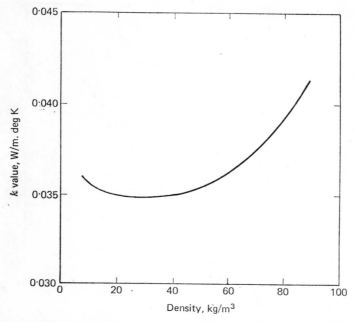

Fig. 12.3 Relationship between the density and k-value of expanded polystyrene at an ambient temperature of 20°C

use it unsupported for thicknesses in excess of 3 cm in this way, as it tends to compress. It is widely used for wall insulation and for the construction of thermally insulated curtain walling in industrialized building methods Expanded polystyrene ceiling tiles are a useful and cheap way of providing thermally insulating and acoustically absorbing ceiling finishes. When expanded polystyrene is used on the inside of a building, it acts as its own water vapour barrier. Expanded polystyrene pellets are used as a loose fill for wall insulation and ceiling insulation.

12.3 Expanded ebonite and rubber

Cellular rubbers can be subdivided into the following groups:

1 *Sponge rubber*, in which the cells are either completely or partially connected. This material is made from masticated raw rubber.

2 *Latex foam rubber* has non-interconnecting cells and is made directly from natural latex foam.

3 *Expanded rubber* is made from masticated raw rubber and has non-interconnecting cells. A material which is made in a somewhat similar way is expanded ebonite.

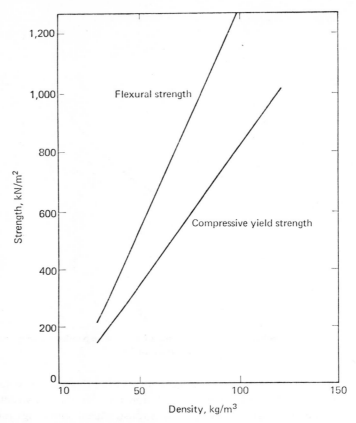

Fig. 12.4   Flexural and compressive yield strengths of expanded polystyrene as
a function of foam density

EXPANDED EBONITE

This is produced from rubber which contains 30% or more of sulphur. Nitrogen is liberated in the material during its manufacture and forms non-interconnecting pores. In the fresh state, expanded ebonite contains bubbles of nitrogen and traces of hydrogen sulphide. Both these gases gradually diffuse through the cells and are replaced by air. It takes from 2—9 months for every trace of hydrogen sulphide odour to disappear. The density of the material is usually 65 kg/m³ and at this density the material has a thermal conductivity somewhat below that of other high-void cellular plastics. Expanded ebonite can only be used at temperatures below 50°C and the cells start collapsing at 75°C. Although the material is basically inflammable, it is self-extinguishing due to the fact that, when it burns, gases are given off which smother the flames. It is classified as BS 476:1932 Class 3 'Material with low

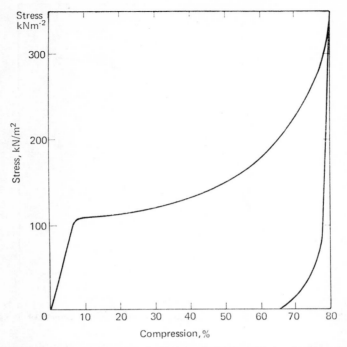

Fig. 12.5   Stress versus compression hysteresis for expanded polystyrene board
with a density of 16 kg/m³

inflammability'. It is interesting to note that the ignition temperature of
expanded ebonite is as high as 450°C, the same as low-temperature coke.

Properties of expanded ebonite
The material has one of the lowest of all water absorption figures, amounting
to 1·5% by volume when immersed in water for 7 days, or about 0·15% after
15 days' exposure to 100% relative humidity.

The various strengths of expanded ebonite differ markedly with the temperature at which the tests are carried out. At 20°C the compressive strength
of expanded ebonite with a density of 65 kg/m³ is equal to 280 kN/m² and its
tensile strength is about 420 kN/m² (see Fig. 12.7). The shear strength amounts
to 120 kN/m² and the modulus of elasticity to 17·5 MN/m². The un-notched
Izod impact strength of a sample tested according to BS 903 Part E 6 has
been found to be 0·12 kg.m. Many of these figures improve as the expanded
ebonite is cooled and therefore this material is excellent for use in cold storage
construction. This is enhanced by its very low water vapour transmission
figure, which is only 1 g/m².hr.bar per inch (25·4 mm) thickness, the
lowest value applying to any of the common cellular plastic materials. The
material is only slightly affected by chemicals, being attacked by aromatic

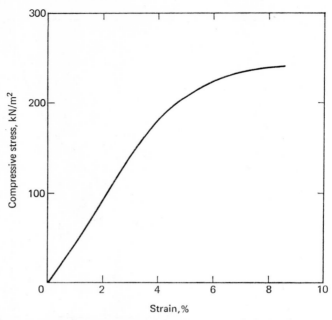

Fig. 12.6 Stress/strain curve of expanded polystyrene with a density of 35 kg/m³

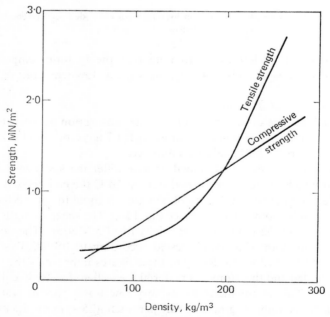

Fig. 12.7 Relationship between the density and compressive and tensile strengths for high-density foamed ebonites

hydrocarbons, chlorohydrocarbons and esters, acetone and similar solvents, but is quite inert to most inorganic materials. It shows no tendency to age and has excellent electrical insulation properties. The material is mainly used in the form of panels which can be glued to walls. Both hot bitumen and cold plastic glues are usually employed for fixing.

12.4 Phenolic foams

The phenolic foams are prepared from phenol formaldehyde resols which contain an active methylol group ($-CH_2OH$). To induce the foaming reaction a blowing agent and various modifiers are added. On the addition of an acid hardener, gas is liberated, resulting in the rapid formation of a liquid froth which has between four and five times the original volume of the resin. The heat developed in the further reaction causes additional expansion and setting of the foamed mass. The time taken for foam with a density of 56 kg/m³ to be formed is approximately 5 min. Different types of phenolic resins are used when foams of different densities are required and the usual range of densities produced is between 8 and 160 kg/m³. The exothermic reaction which causes the foam to set liberates steam, as the temperature within the structure exceeds 100°C. During cooling there is a linear shrinkage of 2% and a further shrinkage of 3% takes place during setting. The foam can be made by a simple batch process, or alternatively it is made in a machine similar to that used for making polyurethane foam.

Properties of phenolic foams
The material has a pink to dark brown appearance and consists of approximately 50% interconnecting cells when made to the standard specification of 56 kg/m³ density. The average cell size is 2·5 mm in diameter.

The strength figures given for 56 kg/m³ material are as follows (see also Fig. 12.8):

Compressive strength	230 kN/m²
Shear strength	145 kN/m²
Tensile strength	130 kN/m²

Phenolic foams can be used at much higher ambient temperatures than most other types of plastic foam and a maximum operating temperature of 130°C is commonly quoted.

On the other hand, phenolic foams are very poor from the point of view of water absorption. After 7 days' immersion in water at 18°C the percentage moisture absorbed amounts to 60% by volume. The water vapour diffusance figure is also very high and amounts to 0·5 kg/m². hr. bar for board 1 in thick (25·4 mm), which is roughly ten times the value which applies to even the lowest density grade of expanded polystyrene. On the other hand, foamed phenol formaldehyde has about the best resistance to ignition of any foamed plastics. Phenolic foam was tested by the Fire Research Station and was found to obey the conditions of 'very low flame spread' (BS 176 Part 1 1953 Class 1). When the foam is exposed to a temperature of about 127°C for a long time, there is a tendency for it to become brittle and to darken. Phenolic foam has a very low coefficient of thermal expansion, and the figure quoted for normal materials is $1 \cdot 1 \times 10^{-5}$ degC^{-1}.

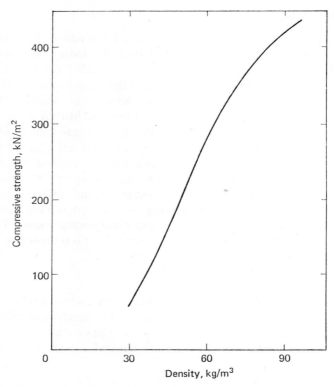

Fig. 12.8   Relationship between the density and compressive strength for phenolic foam

One of the main uses in the building industry for this material is in the manufacture of structural building panels filled with a foam of an approximate density of 32 kg/m^3. Such panels are usually made from mild steel and measure approximately 330 cm $\times$ 120 cm $\times$ 10 cm. Another application is in the insulation of roofs constructed from reinforced concrete. Blocks of phenolic foam with a maximum density of 48 kg/m^3 are laid upon a bituminous matrix applied over the concrete and are then surfaced with layers of bitumen and felt. The foam is not attacked by the hot bitumen, and also has sufficient compressive strength to enable men to walk upon it. Other building units made from the material are sandwich units composed of a phenolic foam core and clad on each side by glass-fibre-reinforced epoxide sheeting.

Such panels are very light in weight, have a good strength and possess excellent thermal insulation properties.

12.5 Foamed polyethylene and related materials

This is one of the most recent developments of all in the field of expanded plastics. Polyethylene is first of all modified by exposure to ionizing radiation in a Van de Graaf accelerator, which causes cross-linking to take place, thereby strengthening the cell walls sufficiently to prevent them from rupturing during the production of the cellular material. The finely divided polyethylene moulding powder, mixed with a blowing agent, is fed into a plastics extrusion hopper, and the material is extruded continuously as a flexible strip. The material has a density of about 32 kg/m³ and is extremely soft and flexible. Its main use is as a sealing strip to prevent ingression of water between adjoining curtain wall panels, as a flexible bed for roof sections and for similar purposes. Foamed polyethylene has the advantage of immunity from ageing, but it also has a very low temperature limit of use, namely about 50°C.

Related to foamed polyethylene is expanded polypropylene, which can be expanded by a blowing agent such as dichlorotetrafluoroethane, which is introduced at a pressure of 25 bar and is allowed to permeate the polymer for 8 hr at 175°C. A material of density 40 kg/m³ is produced. The material has roughly similar characteristics to blown polyethylene but has somewhat higher rigidity and better temperature resistance.

Chlorosulphonated polyethylene is also produced in the form of a cellular product, but the density of the material is rather higher than that of expanded polythene or polypropylene. Chlorosulphonated polyethylene is mixed with various plasticizers, magnesium carbonate and dinitroso-pentamethylene tetramine. On gentle heating to 140°C it expands to a foam with closed pores and a density of between 130 and 150 kg/m³, in the form of sheet 16 mm thick. This material is used for such purposes as sealing horizontal joints between concrete and gas concrete blocks. The compression of this strip has been found to be 22% under a load of 7 tonnes/m² and 55·6% under a load of 21 tonnes/m². It is generally impractical to make this foam in thicknesses exceeding about 16 mm.

12.6 Cellular polyurethanes

Polyurethanes are plastics which are made from isocyanates and alcohols, the basic reaction being represented by the following:

$$-R-NCO+HOR'-$$
$$-R-NH-COO-R'-$$

The isocyanates are all very reactive materials and the reactions are both started and controlled by initiators, as described in Chapter 10.

Polyurethanes have four main uses:

1 The production of rubbers.
2 The production of surface coatings, described in Chapter 9.
3 The production of rigid foams.
4 The production of flexible foams.

RIGID POLYURETHANE FOAMS

Rigid polyurethane foam is used in the building construction field for the filling of hollow wall and roof panel units, which can then be used as highly insulating external wall panels and for *in situ* applied floor insulation, when the polyurethane is simply sprayed between a set of flooring joists resting directly upon a concrete floor. It is widely used for spray application upon internal wall surfaces, when it constitutes a most useful and easily applied thermal insulation layer.

Rigid foams are made by mixing suitable resins, isocyanates, catalyst, water and emulsifier together using a continuous foaming machine.

Typical reactions such as:

$$R-NCO+H_2O \rightarrow RNH_2+CO_2$$
$$R-NCO+RNH_2 \rightarrow R-NH-CO-NH-R$$

take place and are strongly exothermic. The carbon dioxide liberated during the reaction is the foaming agent, and the reaction can be controlled accurately by suitable formulation. A very large number of different isocyanates are used in the production of rigid foams, each with their own special characteristics.

Several polyurethane foaming machines are on the market, which carefully meter the various liquids together, mix them at the nozzle and eject them there. During the actual setting process the rigid foam heats up very considerably, and temperatures of up to 130—140°C have been measured in the centres of some foams. The viscosity of the foam when expelled from the machine is about 3 P at ambient temperatures. It can be applied at the rate of over 2 m²/min in a layer 2 cm thick when used for the insulation of inside surfaces, the foam consisting of closed cells of a diameter of 0·1 mm. The physical properties of polyurethane foam vary considerably with the type of resin and polyol used, as well as the amount of plasticizer employed.

Properties of polyurethane foams
Shell Chemicals Limited give the following properties from their formulations in which Caradol T560R, a blend of polyether polyol, catalysts, emulsifying agents, halogenated organophosphate such as Caradate 30:

$$OCN-C_6H_4-CH_2-C_6H_4-NCO$$

employing trichlorofluoromethane CCl_3F as a blowing agent, are used:

Caradol T560R	100 parts
Caradate 30	120 parts
CCl_3F	40 parts

Using a Viking Mark V dispensing gun, which operates with an air pressure of 5·5 bar, a throughput of about 3 kg/min is obtained. The cream time is given as 25—50 sec, full-time rise as 65—105 sec and tack-free time as 75—200 sec. The density of the final foam is 34 kg/m³ and the percentage closed cells is found to be 89. The thermal conductivity is measured at 0·187 W/m.degC and the water vapour permeability at 0·10 kg/m².hr.bar for 1 in thick board (25·4 mm). The compressive strength of the material is given as 145 kN/m² and the flexural strength at 340 kN/m². The percentage volume change which takes place when a sample is exposed to 100% relative humidity for 48 hr at a temperature of 70% is found to be 4%.

The material is inflammable to some extent but self-extinguishing. Some flame-resistant grades are, however, available. When polyurethane foam is exposed to water at 18°C for 7 days, the percentage volume absorption for trichlorofluoromethane-blown varieties is 5·5, which is somewhat higher than for the normal carbon-dioxide-blown type.

ICI Limited give some figures with regard to their polyurethane foams. A typical example would be their materials made from a mixture of di-isocyanato-diphenyl methane and hexane triol with CO_2-filled pores. The ratio of polyol to triol is 20/80 and the density 33·5 kg/m³. Compression set unheated is 14·5% and after heating for 4 hr at 120°C it sets another 4·6%. Tear strength is equal to 0·77 kg/cm and tensile strength 150 kN/m². Elongation is 290% with a resilience figure of 31%.

Polyurethane foams are stable up to 90°C, but some formulations, which tend to be expensive, can be used up to 180°C. In general, the advantage of the Freon-blown polyurethanes is a lower thermal conductivity than that of carbon-dioxide-blown polyurethane, but at the expense of dimensional stability and resistance to water vapour transmission.

FLEXIBLE FOAMS

These are used widely in the building industry for numerous sealing purposes, for the absorption of sound, for insulating pipelines and water tanks, and for making underlays for carpets and other floor coverings. Mats made from flexible polyurethane foam are also often used for 'floating floor' construction to minimize sound transmission between storeys.

Flexible foams are made by the production of an emulsion of a polyether such as the ICI Daltocel T56, a mixture of 2,4- and 2,6-tolylene di-isocyanates and various catalysts. The amount of carbon dioxide in the foam is usually determined by the proportion of water in the mix. For example, 3% water produces a foam of density 30 kg/m³, while an increase to 4·5% water reduces the density to 23 kg/m³.

Properties of flexible foam

The following data are given by ICI Limited with reference to their Daltocel T56–Supracec EN formulation No. 1:

Density	30 kg/m³
Indentation hardness	21 kg according to BS 3379:1961
Hardness index change	Less than 25%
Tensile strength	120 kN/m²
Elongation at break	225%
Compression set after 22 hr at 70°C:	
50% compression	3%
90% compression	8%
Resilience according to ADTM 1564/59 T	42%
Flex fatigue—loss in hardness	9%

Heat ageing is at 140°C and the retained tensile strength after 2 days is 88%, and after 7 days, 69%. Humidity ageing consists of 16 hours' exposure to steam at 105°C and produces a retained tensile strength of 87%. Lower densities, such as formulation No. 4, which only weighs 23 kg/m³, tend to have a lower flex fatigue, but in other respects they behave practically similar to the denser specimens.

12.7 Urea-formaldehyde foam

This material is extremely popular for insulating existing brick or timber cavity walls. The reason for this is the ease with which the urea-formaldehyde foam can be injected into such cavities through small holes on the inside or outside of the wall, and that, unlike polyurethane foam, it sets without any appreciable heat change.

The urea-formaldehyde has first to be produced in the half-way stage. Sixty parts of the urea are mixed with 200 parts of 30% aqueous formaldehyde, and neutralized with sodium hydroxide to pH 7. The mixture is boiled for 10 min and 0·26 parts of acetic acid are added, followed by further heating. This liquid is transported to the building site and placed into one of the two cylinders of the injection machine. The second cylinder contains a mixture of initiator and a detergent. Thin streams of the two liquids are whipped together into a foam and injected through a nozzle into the wall. There it sets to a highly porous solid foam, which dries out in due course to a water-repellant mass of density 5—8 kg/m³ and of the consistency of cotton wool. The pores are very small and completely interconnecting so that the material has virtually no resistance to water vapour. Urea-formaldehyde foam can be made so as to contain a mere 0·7% of solid material and 99·3% air. Its compressive strength and tensile strength are virtually zero, and in consequence the material must at all times be duly supported on all sides by other materials. Some of the older samples used to exhibit fissuring, but this has now been virtually eliminated.

The material in its final state is unaffected by dilute acids, alkalis, oils or solvents. It also does not rot or show any sign of ageing. It is incombustible and simply shrivels up when exposed to high temperatures. The material is greasy in nature, so that, while it presents no barrier to water vapour, it tends to repel liquid water by capillary action.

The comparative cheapness of the material, ease of handling, and the fact that it permits walls to breathe have all contributed in the popularity of the material for *in situ* wall insulation. However, when it is used for this purpose, it becomes necessary to instal a water vapour barrier on the inside of the foam, otherwise it is found that condensation may arise within the body of the foam due to the ease with which water vapour passes through the open pores of the material.

Water absorption after 7 days' immersion equals 2·5% by volume.

Literature Sources and Suggested Further Reading

1. COOPER, A., 'Properties of Cellular Polymers, *Trans. Plastics Inst.*, Vol. 26, No. 65, July (1958)
2. DIAMANT, R. M. E., *Insulation of Buildings*, Iliffe, London (1965)
3. FERRIGNO, T. H., *Rigid Plastics Foams*, Reinhold, New York (1963)
4. ICI LIMITED, Numerous Technical Papers and Brochures
5. MONSANTO CHEMICAL LIMITED, Technical Brochures
6. SAUNDERS, J. H., and FRISCH, K. C., *Polyurethanes*, Interscience, New York (1964)
7. SHELL CHEMICALS LIMITED, *Expanded Polystyrene in Building* (1968)

The literature references given in Chapters 10 and 11 are also frequently applicable.

Index